ADDISON WESLEY

Math
Makes Sense

2

Author Team

Carole Saundry	Sharon Jeroski
Heather Spencer	Michelle Jackson
Maureen Dockendorf	Sandra Ball
Maggie Martin Connell	Jill Norman
Linden Gray	Susan Green

PEARSON

Addison
Wesley

Elementary Math Team Leader
Anne-Marie Scullion

Publisher
Claire Burnett

Publishing Team
Enid Haley
Lesley Haynes
Tricia Carmichael
Lynn Pereira
Rosalyn Steiner
Ellen Davidson
Sarah Mawson
Eileen Pyne-Rudzik
Stephanie Cox
Kaari Turk
Judy Wilson
Nicole Argyropoulos

Product Manager
Nishaant Sanghavi

Photo Research
Karen Hunter

Design
Word & Image Design Studio Inc.

ISBN 0-321-11814-6

Printed and bound in Canada

2 3 4 5 TCP 09 08 07 06 05

PEARSON

Addison
Wesley

Acknowledgments
The publisher wished to thank the following sources for photographs, illustrations, and other materials used in this text. Care has been taken to determine and locate ownership of copyright material in this book. We will gladly receive information enabling us to rectify any errors or omissions in credits.

Cover
Cover illustration by Marisol Sarrazin

Illustrations
Kasia Charko, p. 253
Virginie Faucher pp. 141, 214–234
Marie-Claude Favreau, pp. 1–12, 75–86, 177–188, 269–280
Eugenie Fernandes, p. 193
Joanne Fitzgerald pp. 88–90, 236–252
Leanne Franson, p. 153
Linda Hendry, pp. 115–132
Tina Holdcroft, pp. 53–61, 63, 66–71, 73, 74, 154–176
Vesna Krstanovic pp. 27–29, 31–49, 52, 194–212, 217
André Labrie, pp. 13–26
Bernadette Lau, pp. 50, 51, 200
Paul McCusker, pp. 64, 65, 104, 107, 151
Marc Mongeau, pp. 213, 235
Allan Moon, pp. 11, 30, 43, 44, 61–63, 72, 74, 118, 119, 121, 122, 135, 137–140, 144–149, 208, 209
Scott Ritchie, pp. 91–94, 96–103, 105–109, 111, 112, 114, 190–192, 254–268, 281–284
Bill Slavin, pp. 133–152
Pat Stephens, p. 95
Neil Stewart, pp. <Math at Home tech art>

Photography
Ray Boudreau pp. 87–90, 189–190, 192, 281

Contents

Program Consultants and Advisers

Program Consultants

Craig Featherstone
Maggie Martin Connell
Trevor Brown

Assessment Consultant
Sharon Jeroski

Primary Mathematics and Literacy Consultant
Pat Dickinson

Elementary Mathematics Adviser
John A. Van de Walle

British Columbia Early Numeracy Advisor
Carole Saundry

Ontario Early Math Strategy Adviser
Ruth Dawson

Program Advisers

Pearson Education thanks its Program Advisers, who helped shape the vision for *Addison Wesley Mathematics Makes Sense* through discussions and reviews of prototype materials and manuscript.

Anthony Azzopardi
Bob Belcher
Judy Blake
Steve Cairns
Daryl Chichak
Lynda Colgan
Marg Craig
Jennifer Gardner
Florence Glanfield
Pamela Hagen
Dennis Hamaguchi
Angie Harding
Peggy Hill

Auriana Kowalchuk
Gordon Li
Werner Liedtke
Jodi Mackie
Kristi Manuel
Lois Marchand
Cathy Molinski
Bill Nimigon
Eileen Phillips
Evelyn Sawicki
Shannon Sharp
Lynn Strangway
Mignonne Wood

Program Reviewers

Field Testers

Pearson Education thanks the teachers and students who field-tested *Addison Wesley Math Makes Sense 2* prior to publication. Their feedback and constructive recommendations have been most valuable in helping us to develop a quality mathematics program.

Aboriginal Content Reviewers

Early Childhood and School Services Division, Department of Education, Culture, and Employment, Government of Northwest Territories:

Steven Daniel, Coordinator, Mathematics, Science, and Secondary Education
Liz Fowler, Coordinator, Culture-Based Education
Margaret Erasmus, Coordinator, Aboriginal Languages

Grade 2 Reviewers

Anne Boyd
School District 72
(Campbell River), BC

Bob Belcher
Sooke School District, BC

Judy Blake
School District 44 (North Vancouver), BC

Trevor Brown
Course Director,
Mathematics Education,
OISE/UT, ON

Ralph Connelly
Professor Emeritus
Brock University, ON

Marg Craig
Independent Mathematics
Consultant, ON

Ruth Dawson
Coordinator, Halton District
School Board, ON

Lorelei Gibeau
Edmonton Catholic Separate
School District, AB

Werner Liedtke
University of Victoria, BC

Lois Marchand
Independent Consultant, AB

Livia Paradis
Edmonton Catholic School
Board, AB

Gillian Parsons
Elementary Program
Co-ordinator, Brant
Haldimand-Norfolk Catholic
School Board, ON

Lynn Strangway
Toronto District School
Board, ON

Roz Thomson
Halton District School
Board, ON

School Begins

"It's time for school," Cam's grandma said.
"Open your eyes. Get out of bed.
I'm coming to school today with you,
because you're starting somewhere new."

"I'm scared," Cam said, then chewed his toast.
He hated changing schools the most.
"I won't have friends."
Then Grandma sighed.
"You'll make *new* friends, Cam," she replied.

The teacher greeted them at the door,
and welcomed Cam to Classroom 4.
The chairs and tables were tightly fit.
Cam looked around. "Where should I sit?"

The teacher pointed. "There's a chair.
You can sit with those children there.
Your grandma can sit at that table, too,
if she'd like to stay awhile with you."

The teacher asked Grandma, "Can you help today?
We're learning math games we can play.
There are materials to share and rules to learn,
like when to move and take a turn."

Cam's group listened to the teacher explain.
She repeated some of the rules again.
She said, "Do you have questions? Raise your hand."
But the class said, "No, we understand."

At recess, Grandma had to go.
She said, "This is the best Grade 2 I know!"
She waved at all her new young friends,
who called, "Come back, and help again!"

About the Story

The story was read in class to prepare for a Mathematics Investigation activity. Children played a variety of mathematical games and created addition and subtraction stories. The Investigation provided opportunities for the teacher to learn about children's mathematical understanding and skills as they begin a new school year.

Talk about It Together

• How did Cam feel about his first day of school?
• What do you think Cam's math class will be like? Why?
• What did Cam's teacher do to help the children work together?
• How is Cam's classroom the same as your classroom? How is it different?

At the Library

Ask your local librarian about books with math-related themes that are appropriate for Grade 2 readers.

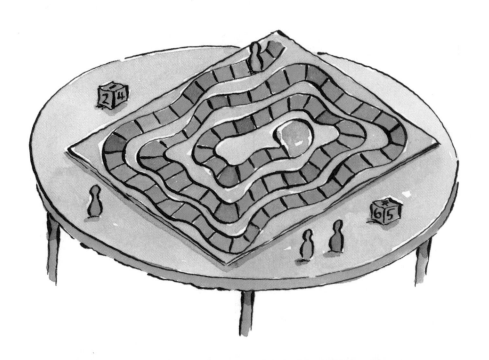

Grandma Helps

Make a number story about the picture.
Use pictures, numbers, or words to tell your story.

9

Race to 100!

Roll 2 number cubes.
Add the numbers.
Take that number of Snap Cubes.

Every time you have 10 Snap Cubes,
put them together to make a 10-stick.
Stop when you have 10 sticks. They will total 100 cubes!

Estimate. How many rolls will you need to get 100 cubes?
Show your thinking in pictures, numbers, or words.

Keep track. Make a ✔ for each roll.
When you reach 100, count your ✔.

How many rolls altogether? _____

Building Challenge!

Use a spinner.
Try to build the tallest structure you can.
You can spin 10 times. Which spinner will you choose?

Show your thinking in pictures, numbers, or words.

Spin 10 times.
Each time you spin, take the object you landed on.
Tell how many of each you have.

3-D Object	How Many I Have	How Many I Used
cone		
cylinder		
sphere		
cube		
prism		

Use your 10-stick to measure your structure.

How tall is it? _____ 10-sticks

A Special Game

Make your own math game.
You can use spinners, number cubes,
counters, solids, or anything you like.

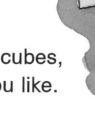

Tell how to play your game.

What do you need to play your game?

Write the rules to your game.

Teach other children how to play your game.

Sorting and Patterning

Dear Family,

Your child is starting a unit in mathematics on sorting
and patterning.

The Learning Goals for this unit are to

- Sort objects according to attributes, such as colour,
  size, and shape.
- Describe, extend, and draw patterns.
- Talk about a pattern rule.
- Use two attributes to make a pattern.

You can help your child achieve these goals
by doing the Home Connection activities
suggested at the bottom of selected pages.

Clean-Up Time

Help the carpenter clean up her bench.

Focus | Children decide how to sort the items into the bins. They colour each item to match the bin where that item belongs. There are multiple correct answers.

Name: _____   Date: _____

Find the Rule

Write labels in the boxes to tell how the clothing is sorted.

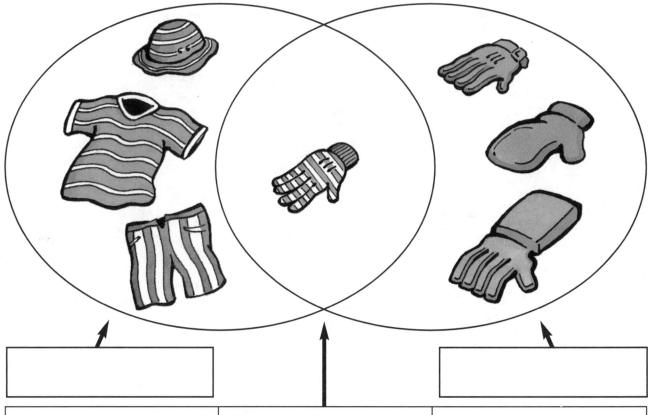

⌃	⌃	⌃
Draw something else that belongs in the first sorting circle.	Draw something else that belongs in both circles.	Draw something else that belongs in the second sorting circle.

FOCUS Children identify the sorting rule and label the sorting circles. They draw other items that belong in each group.

HOME CONNECTION
Ask your child: "Why do we sort laundry before washing it and putting it away? What other things do we sort?"

Make Your Own Patterns

Use Attribute Blocks.

Make a pattern with 2 changing attributes. Draw it here.

[blank box]

Circle the pattern core.

List the changing attributes. _____ _____

Use the same 2 attributes.

Make a different pattern. Draw it here.

[blank box]

Circle the pattern core.

How are the patterns the same?

_____

How are they different?

_____

Focus | Children make and record patterns with two changing attributes.

Unit I, Lesson 2: Make a Pattern   **17**

What Is Changing?

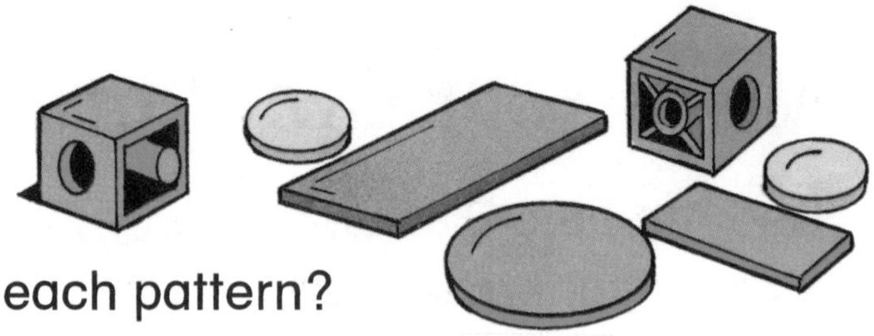

Circle the core in each pattern.
What 2 attributes are changing in each pattern?

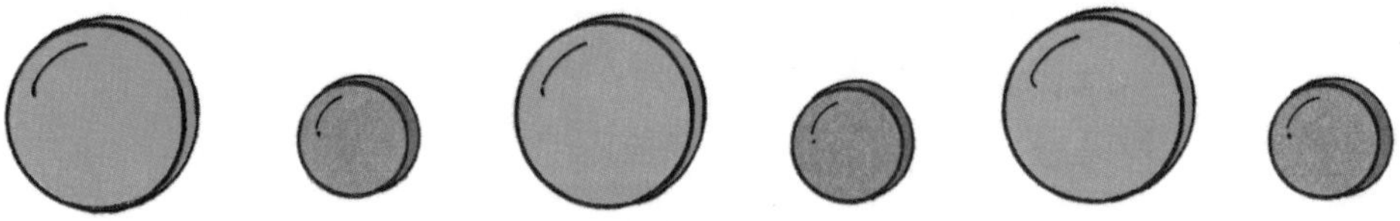

_____   _____

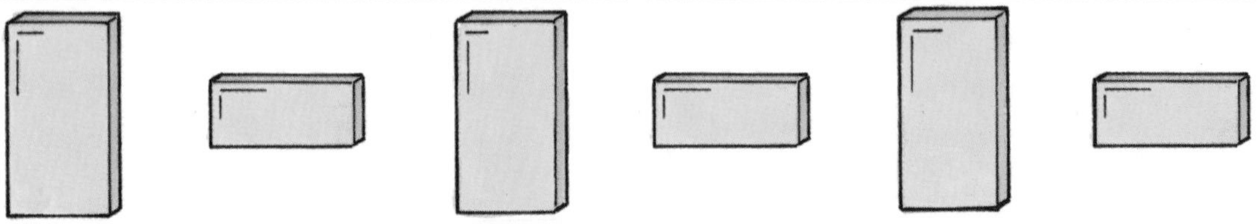

_____   _____

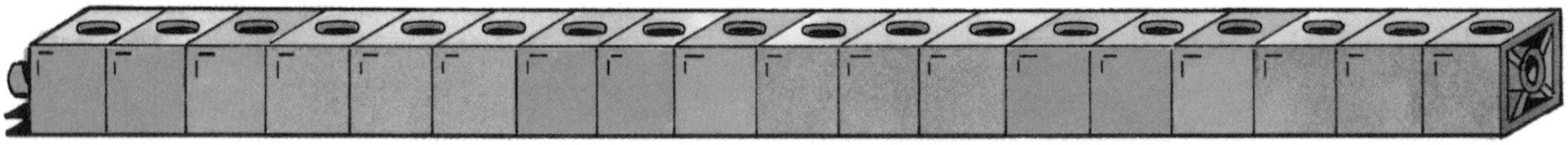

_____   _____

Make a pattern. Draw it here.

Name: _____   Date: _____

Making Patterns

Make a pattern. Use 6 △ and 9 ☐ .

```


```

Look at a friend's pattern. How are the patterns the same?

_____

How are the patterns different?

_____

How did you know how to arrange the objects?

_____

_____

HOME CONNECTION
Have your child use large red and small green circles to make a pattern in two different ways. Ask: "How are the patterns the same? How are they different?"

Name: _____   Date: _____

Make the Same Pattern

Look at the Snap Cube pattern.

Here is another way to show the pattern.

Draw two other ways to show the pattern.

```



```

```



```

Name: _____   Date: _____

Picture Patterns

Draw pictures to show the patterns.

tall, short; tall, short; tall, short

[ ]

red, blue, green; red, blue, green; red, blue, green

[ ]

Use words to write your own pattern.

_____

_____

Ask a friend to draw your pattern.

[ ]

HOME CONNECTION
Look around your home with your child for patterns. Ask your child
to draw a picture to represent each pattern.

Focus | Children use picture patterns
to represent word patterns.

Name: _____   Date: _____

Name That Pattern

Use words to describe the pattern core.

_____

Use letters to record the pattern.

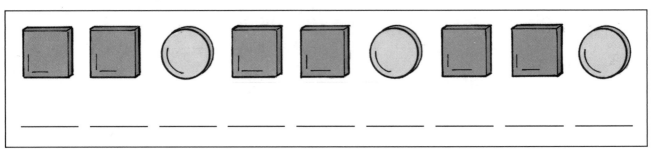

__ __ __ __ __ __ __ __ __

Use numbers to record the pattern.

__ __ __ __ __ __ __ __ __

Draw a pattern. Ask a friend to record the pattern in another way.

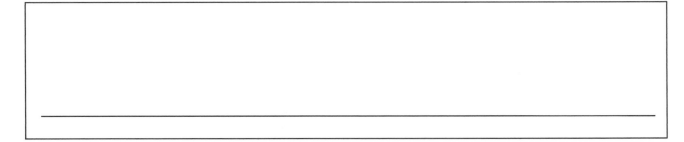

Name: _____ Date: _____

Use All the Beads

There are 9 red, 3 green,
and 6 blue beads.

Make a pattern.
Use all the beads.

Show your thinking in pictures, numbers, or words.

Name: _____   Date: _____

Bead Pattern

There are 3 blue, 6 red,
and 6 green beads.

Make a pattern.
Use all the beads.

Show your thinking in pictures, numbers, or words.

Focus | Children use all the beads to make a pattern.
There are multiple correct answers.

Name: _____ Date: _____

My Pattern Border

Draw the pattern for your border and describe it.

_____

_____

HOME CONNECTION
With your child, make a pattern using crayons, markers, stickers, pencils, or other favourite objects. Change the order of objects and make another pattern.

Name: _____ Date: _____

My Journal

Tell what you learned about using 2 attributes to sort.

Tell what you learned about using 2 changing attributes to make a pattern.

FocuS | Children reflect on what they have learned about sorting and patterning.

26 Unit I, Lesson 5: Show What You KnowCopyright © 2005 Pearson Education Canada Inc. Not to be copied.

Number Relationships

Focus | Children talk about the picture and identify numbers of objects.

Name: _____ Date: _____

Dear Family,

This unit will focus on deepening your child's understanding of number relationships, counting, and place value.

The Learning Goals for this unit are to

- Read and print number words to 20.
- Build numbers with concrete materials.
- Estimate the number of objects and check by counting.
- Count forward to 100 and backward from 20 using a number line, a 100-chart, and a calculator.
- Count by 1s, 2s, 5s, 10s, and 25s.
- Develop strategies for adding and subtracting.

You can help your child achieve these goals by doing the Home Connection activities suggested at the bottom of selected pages.

How Many?

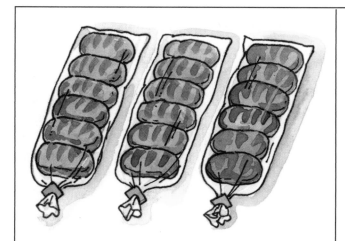

How many rolls? _____

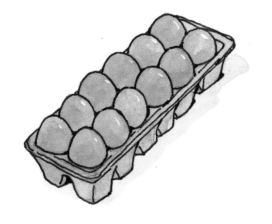

How many eggs? _____

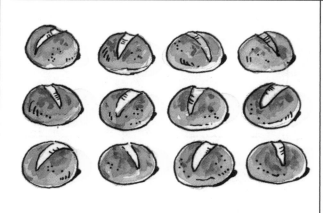

How many buns? _____

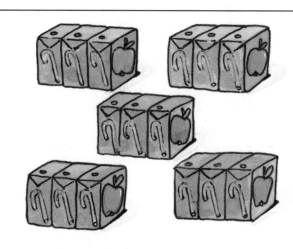

How many boxes? _____

What groups of items show the same number?

_____

How do you know?

_____

Numbers to 20

Draw counters to show how many. Record the numerals.

twelve

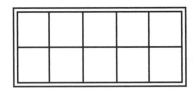 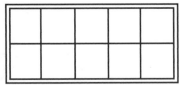

_____ is 10 and _____.

nineteen

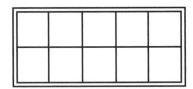 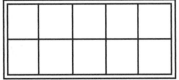

_____ is 10 and _____.

fourteen

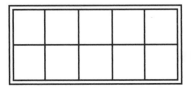 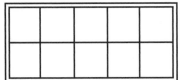

_____ is 10 and _____.

Print the number words. Record the numerals.

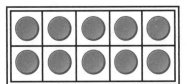 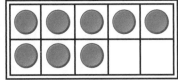

_____

_____ is 10 and _____.

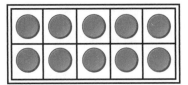 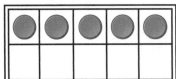

_____

_____ is 10 and _____.

HOME CONNECTION
Have your child explain how to use a ten-frame. Ask: "How would you use a ten-frame to show 11? To show 17?"

Counting Two Ways

Spill the objects. Estimate the number. _____
You will count all the objects.

As you begin counting, look back at your estimate.
If you want, make a new estimate. _____

Tell how you counted. Use pictures, numbers, or words.

Spill the objects again. Count them another way.
Tell how you counted. Use pictures, numbers, or words.

Name: _____ Date: _____

Count the Buttons

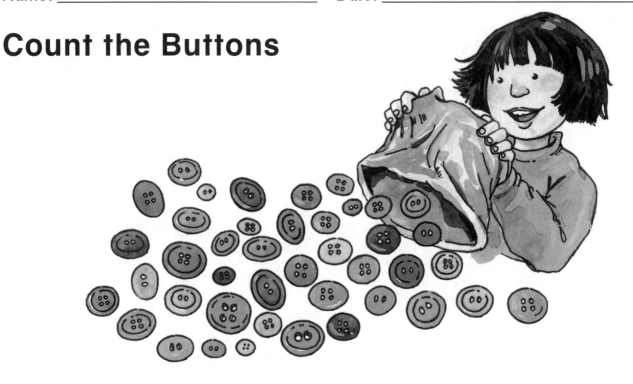

How many buttons are there? _____
Tell how you counted. Use pictures, numbers, or words.

```




```

What other way could you count the buttons?

_____

_____

 HOME CONNECTION
Gather a collection of about 40 small objects for your child to count,
such as raisins or pennies. Ask your child to count the collection by
grouping the objects in different ways.

Focus | Children count a collection
       using a strategy of their choice.

What Is Missing?

Write the missing numerals on the number lines.

8 ☐ 10 ☐ 12 13 14 ☐ ☐ 17 18

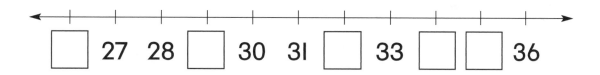

☐ 27 28 ☐ 30 31 ☐ 33 ☐ ☐ 36

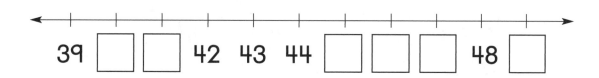

39 ☐ ☐ 42 43 44 ☐ ☐ ☐ 48 ☐

Write numerals on the number line for these number words:
fifteen, eighteen, twenty, twenty-two.

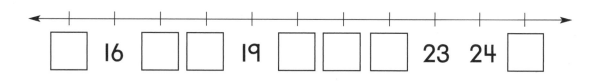

☐ 16 ☐ ☐ 19 ☐ ☐ ☐ 23 24 ☐

What numerals are still missing? _____
Write these numerals on the number line.

HOME CONNECTION
Help your child find some examples where number lines are used,
such as on a thermometer or on a map showing the map scale.

Focus | Children complete partial number
lines by writing missing numerals.

Number Sentences

Write each number sentence.

_____ + _____ = _____   _____ + _____ = _____

_____ − _____ = _____   _____ − _____ = _____

Make your own

_____ ◯ _____ = _____

34 Unit 2, Lesson 4: Number Facts to 18

Name: _____   Date: _____

Garden Problems

Write each number sentence.

There are 9 in a garden.

4 are yellow. The rest are pink.

How many are pink?

_____ ◯ _____ = _____

There were 13 on a tree.
Some fell off.

Now there are 6 on the tree.

How many fell off?

_____ ◯ _____ = _____

There are 15 in the yard.

8  are in a tree.

The rest are on a fence.
How many are on the fence?

_____ ◯ _____ = _____

A puts 7  in a pile.

The squirrel gathers some more.

Now there are 16 .

_____ ◯ _____ = _____

How many more did the gather?

Focus | Children write number sentences to represent addition- and subtraction-story problems.

HOME CONNECTION
Share addition and subtraction story problems about things in your neighbourhood. For example, "There are 15 houses on our street. 9 of them have a garage. How many do not have a garage?"

Snappy Number Sentences

Use Snap Cubes.
Complete the number sentences.

8+5=13 helps me find the answer to 13−8

_____ + _____ = _____   _____ − _____ = _____

_____ + _____ = _____   _____ − _____ = _____

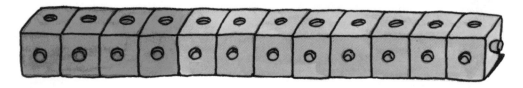

_____ + _____ = _____   _____ − _____ = _____

_____ + _____ = _____   _____ − _____ = _____

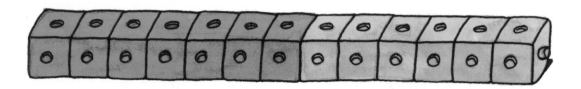

_____ + _____ = _____   _____ − _____ = _____

_____ + _____ = _____   _____ − _____ = _____

How does knowing 7 + 6 help when finding 13 − 6?

_____

Focus | Children write addition and subtraction sentences to describe arrangements of Snap Cubes. They explain how knowing an addition fact helps find the answer to a subtraction fact.

Name: _____   Date: _____

Add or Subtract

I made a group of 10 and then counted on

$$\begin{array}{r} 15 \\ +\ 3 \\ \hline \end{array}$$
$$\begin{array}{r} 9 \\ +\ 7 \\ \hline \end{array}$$
$$\begin{array}{r} 11 \\ -\ 2 \\ \hline \end{array}$$
$$\begin{array}{r} 14 \\ -\ 6 \\ \hline \end{array}$$

$13 + 2 =$ _____ $13 - 5 =$ _____ $17 - 1 =$ _____ $15 - 8 =$ _____

Choose a question. Tell about the strategy you used to solve it.
Use pictures, numbers, or words.

Find the missing numbers. Use counters to model your answer.

$15 - \boxed{} = 6$ $17 - \boxed{} = 8$ $13 - \boxed{} = 6$ $15 - \boxed{} = 7$

HOME CONNECTION

Have your child build a set of 11 to 15 pennies and then add 1, 2, or 3 to that number. Have your child count on from that number to get the total. Repeat the activity for subtraction.

Focus | Children use strategies to complete addition and subtraction sentences.

Seeing Doubles

Finish each picture to show a double.
Write the addition sentence.

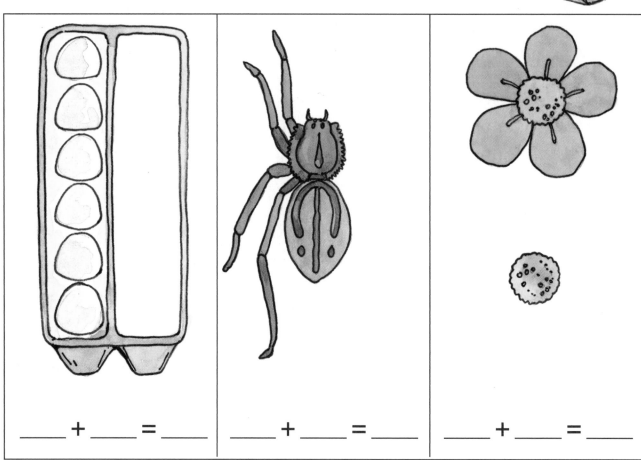

___ + ___ = ___   ___ + ___ = ___   ___ + ___ = ___

Circle the numeral that is not the
answer for a doubles addition story.

| 4 | 8 | 9 | 10 | 16 |

How do you know?

_____

_____

_____

Focus | Children draw to show doubles and record the doubles facts.

38 Unit 2, Lesson 6: Doubles and Near Doubles

Using Doubles

Write 2 doubles facts that can help you find the answers.

6 + 7 = ___	4 + 5 = ___	3 + 2 = ___
___ + ___ = ___	___ + ___ = ___	___ + ___ = ___
___ + ___ = ___	___ + ___ = ___	___ + ___ = ___
7 + 8 = ___	4 + 3 = ___	9 + 8 = ___
___ + ___ = ___	___ + ___ = ___	___ + ___ = ___
___ + ___ = ___	___ + ___ = ___	___ + ___ = ___

HOME CONNECTION
Ask your child to tell a number story using a double or near doubles.

Focus | Children use doubles facts to find answers to near doubles.

Unit 2, Lesson 6: Doubles and Near Doubles **39**

About How Many?

Spill the counters.
Make one group of 10.

Estimate the total number. It is about _____.

Make groups of 10s to show how you counted.
Use pictures, numbers, or words.

_____ groups of 10s and _____ left over   _____ in all

Focus | Children use a group of 10 to estimate
a total, then count groups of 10s and
leftover 1s to determine the total.

Full of Beans

Take a handful of beans with both hands.
Estimate the number.
Think of grouping by 10s.

Circle your estimate.   ⟶ more than 50

⟶ fewer than 50

Make one group of 10.
Change your estimate if you wish. _____

Place the beans in the ten-frames.
Colour to show your work.

_____ groups of 10s and _____ left over   _____ in all

Counting by 10s

Circle groups of 10 ants.
Record the numbers.

_____ groups of 10s and _____ left over _____ in all

Rouda used ten-frames to organize her sticker collection.
How many stickers does she have?

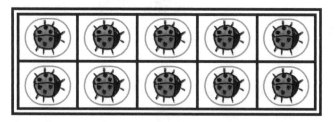

How can grouping by 10s help you with counting?

_____

_____

Be a Number Detective

Each piece of a 100-chart is missing some numerals.
Look for clues in the numbers to help you fill in the empty spaces.

22	23	24		26
	33	34	35	36
42		44	45	46

	52		54	55	
61	62	63		65	66
71	72		74		76

61		63	64	65
71	72	73		
	82		84	

	76		78	79	
	86	87	88		90
95	96		98		100

Page 47 fell out of a book.
How would you tell a friend where it belongs?

_____

HOME CONNECTION
Show a page number between 50 and 100 from a book. Ask:
"What is the next page number? What was the number on the
page that came before? How do you know?"

Focus | Children fill in the missing numerals on pieces of 100-charts.

Unit 2, Lesson 8: Numbers to 100 **43**

Missing Number Mysteries

Fill in some numerals in this piece of a 100-chart.
Trade books with a friend.
Ask your friend to fill in the empty spaces.
Check your friend's work.

71									
									100

How did you choose the numerals to write in?

_____

_____

_____

How did you know where to put the numerals you chose?

_____

_____

_____

Focus | Children use a piece of a 100-chart to make a missing number puzzle for a friend to complete. Then, they check their friend's work.

100-Chart (101 to 200)

Circle numerals that show a pattern.
What is your pattern?

_____

Colour numerals that show another pattern.
What is your pattern?

_____

101	102	103	104	105	106	107	108	109	110
111	112	113	114	115	116	117	118	119	120
121	122	123	124	125	126	127	128	129	130
131	132	133	134	135	136	137	138	139	140
141	142	143	144	145	146	147	148	149	150
151	152	153	154	155	156	157	158	159	160
161	162	163	164	165	166	167	168	169	170
171	172	173	174	175	176	177	178	179	180
181	182	183	184	185	186	187	188	189	190
191	192	193	194	195	196	197	198	199	200

How are the patterns the same?

_____

How are they different?

_____

Focus | Children record and describe two number patterns on a 100-chart that shows 101 to 200.

Unit 2, Lesson 9: Counting Patterns beyond 100 **45**

Name: _____ Date: _____

Odd and Even Numbers

Colour the even numbers from 50 to 68 red.
Colour the odd numbers from 19 to 37 blue.

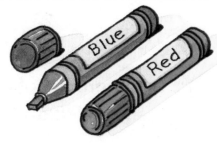

1	2	3	4	5	6	7	8	9	10
11	12	13	14	15	16	17	18	19	20
21	22	23	24	25	26	27	28	29	30
31	32	33	34	35	36	37	38	39	40
41	42	43	44	45	46	47	48	49	50
51	52	53	54	55	56	57	58	59	60
61	62	63	64	65	66	67	68	69	70
71	72	73	74	75	76	77	78	79	80
81	82	83	84	85	86	87	88	89	90
91	92	93	94	95	96	97	98	99	100

What patterns do you see?

_____

_____

_____

HOME CONNECTION

Have your child use the chart on this page to describe some number patterns on a 100-chart.

Focus | Children colour a 100-chart to show odd and even numbers.

Name: _____ Date: _____

Counting Patterns

Look at each list of numbers.
What is the number pattern?

117,	118,	119,	120,	121	counting by _____
130,	140,	150,	160,	170	counting by _____
25,	50,	75,	100,	125	counting by _____
23,	25,	27,	29,	31	counting by _____

Find the patterns.
Write the missing numerals.

46,	48,	50,	___,	___,	___,	58,	___,	___,	64
25,	50,	75,	___,	___,	150,	___,	200,	___,	250
120,	130,	140,	___,	___,	170,	___,	___,	200,	___
95,	100,	105,	___,	115,	___,	___,	130,	___,	___

Focus | Children find and continue counting patterns.

Unit 2, Lesson 9: Counting Patterns beyond 100 **47**

Name: _____ Date: _____

Reaching 41

Will you reach 41 if you begin at 6 and count by 5s?
Show your thinking in pictures, numbers, or words.

Focus | Children use a number pattern to solve a problem.

Name: _____ Date: _____

Reaching 62

Will you reach 62 if you begin at 21 and count by 10s?
Show your thinking in pictures, numbers, or words.

HOME CONNECTION

Have your child explain the clues in this problem
and tell how he or she solved the problem.

Focus | Children use a number pattern to solve a problem.

How Many Paddles?

About how many paddles do you see?

Estimate.   more than 50   fewer than 50

As you begin counting, look back at your estimate.

Change it if you want. _____

Show how you counted. Use pictures, numbers, or words.

There are _____ paddles in all.

Focus | Children estimate and count the number of paddles in a dragon boat race.

Dragon Boat Stories

What is happening on the water?

Tell an addition story.

_____ + _____ = _____

Tell a subtraction story.

_____ – _____ = _____

Focus | Children interpret an illustration and write and solve addition and subtraction story problems.

Unit 2, Lesson II: Show What You Know   **51**

Name: _____ Date: _____

My Journal

Tell what you learned about building numbers.
Use pictures, numbers, or words.

Tell what you learned about large numbers.

Focus | Children reflect on and record what they
learned about number relationships.

UNIT 3

Time, Temperature, and Money

Focus | Children tell a story about the picture and discuss the duration and the order of events.

Dear Family,

In this unit, your child will be learning about time, temperature, and money.

The Learning Goals for this unit are to

- Name and order months and seasons of the year.
- Use ordinals *first* to *thirty-first*.
- Tell time to the quarter-hour on analog and digital clocks.
- Use a thermometer to see if the temperature is rising or falling.
- Count and create money amounts up to $1.00.

You can help your child achieve these goals by doing the Home Connection activities suggested at the bottom of selected pages.

Mixed-Up Apple Times

Look at the drawings.
Number the drawings in the order that they happen.

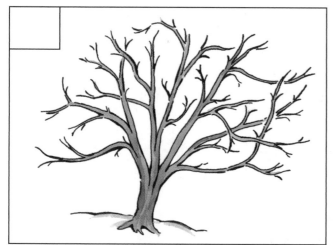

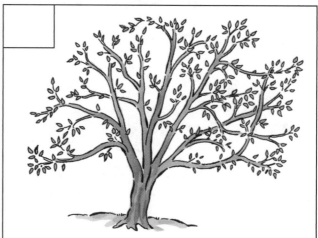

Write a story about the pictures. _____

_____

_____

_____

_____

Counting Time

Count and record the number of swings each activity takes.
Add two more things to do, then count and record the swings.

Activity	Number of Swings
Write your name.	
Say the alphabet.	
Draw a circle and colour it in.	

About Time

Estimate the number of swings each activity will take.
Time each activity. Record the number of swings.

Activity	Estimate	Count
Blink 10 times.		
Draw a face.		
Sing "Row, Row, Row Your Boat."		
Put on your coat.		
Wash your hands.		

HOME CONNECTION
Have your child explain how to use a pendulum
to measure time.

Focus | Children estimate and measure the duration of
activities using a pendulum timer.

How Long Is a Minute?

Count the number of times you can
repeat each activity in one minute.

Activity	Number
Write your whole name.	
Draw a house.	
Make a 5-train with Snap Cubes.	

Focus | Children measure the number of repetitions that fill a minute.

58   Unit 3, Lesson 2: Units of Time

Name: _____   Date: _____

Just a Minute

Write or draw pictures of 3 activities you think
will take one minute.

Time how long each activity takes.

Circle the length of time for each activity.

Activity	
	less than a minute about a minute more than a minute
	less than a minute about a minute more than a minute
	less than a minute about a minute more than a minute

HOME CONNECTION 🖼️
Ask your child to help you prepare a meal. Discuss what
jobs will take one minute. Have your child check with a
timer or a clock that has a second hand.

Focus | Children record activities they think will
take one minute and then measure.

Unit 3, Lesson 2: Units of Time   **59**

Hours, Minutes, Seconds, Days?

Decide whether each event should be
measured in hours, minutes, seconds, or days.

the school day

_____

blowing a bubble

_____

walking a dog

_____

eating dinner

_____

Focus | Children decide what unit of time would be best to measure an event.

Time by 15

Show the time on each clock.

4:15

6:15

9:45

12:45

11:00

10:30

Unit 3, Lesson 3: Telling Time   **61**

Telling Time

Write each time or show it on the clock.

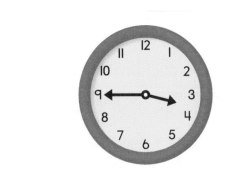
_____

_____

_____

_____

Focus | Children record time to the quarter-hour using analog and digital clocks.

15 Minutes Later

The class is waiting for family day to start.
Family members are invited to come at 11:00.

Beginning at 9:00, someone announces
the time every 15 minutes.

Draw 6 times that are announced.

Focus | Children record quarter-hour intervals on analog clocks.

Name: _____ Date: _____

The Month of October

Look at a calendar for October.
Write the days of the week.
Write the numerals for the month.

How many weeks are in the month of October? _____

How do you know? _____

On what day of the week is the twenty-first? _____

HOME CONNECTION
With your child, discuss special events your family celebrates
during a year. Have your child record these dates on a calendar.

Focus | Children read days and weeks on a calendar.

Calendars

Complete each calendar.

April

Sunday		Tuesday		Thursday		Saturday
		1	2		4	
	7	8		10		12
13		15			18	
			23		25	26
	28					

What day of the week is the fourteenth? _____

the sixteenth? _____

July

Sunday	Monday					Saturday
1		3		5		7
	9	10	11		13	
	16			19		21
22		24		26		
29						

What day of the week is the thirtieth? _____

the eleventh? _____

Focus | Children complete each calendar to show the days of the week and the dates in the month. They interpret the dates given ordinal clues.

Unit 3, Lesson 4: Calendar Time   **65**

Hot or Cold?

Colour to show the temperatures on the thermometers.

at the start   **in hot water**   **in cold water**

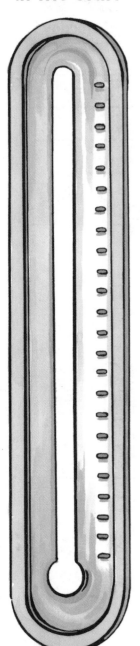   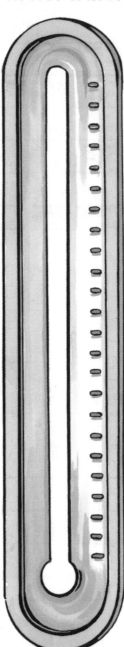   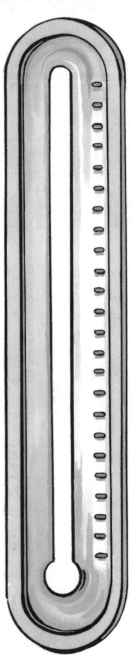

 HOME CONNECTION

Use a thermometer to test the temperature of a warm liquid, and then place the thermometer in a cup of cool water. Repeat with other warm and cool liquids. Ask your child what happens to the thermometer.

Focus | Children record rising and falling temperatures on the thermometer.

For Sale

Draw the coins you could use to buy each of these.

HOME CONNECTION

When shopping with your child, point out price tags on articles under $1.00 and have your child read the amount. Hand your child some change to count to determine if there is enough to buy an item.

What Can I Buy?

You have 95¢.

Circle 3 things you could buy with that exact amount.

Show the coins you would use.

Could you buy 3 other things with that exact amount?

Show your work.

Focus | Children count on and skip count to find the total cost of 3 items.

68 Unit 3, Lesson 6: Making Money Amounts

Counting Coins

You have 6 coins that equal 90¢.
What could the coins be? Draw a picture of your solution.

Focus | Children determine 6 coins that could have a total value of 90¢.

HOME CONNECTION
Have your child explain the solution to the problem.
Ask: "How do you know the coins add up to 90¢?"

What Coins Would You Use?

What 6 coins would you use to make 75¢?
Use pictures, numbers, or words.

How do you know this makes 75¢?

_____

Focus | Children determine what 6 coins they could use to make 75¢.

Name: _____ Date: _____

Sharing Money

Three children find 50¢ in coins.
How can they share the money?
Use pictures, numbers, or words.

On Time

What time did the children leave for school?
Show the time on both clocks.

What time did school end?
Show the time on both clocks.

How is the same as ?

_____

_____

How are they different?

_____

_____

Hotter, Colder

Show the temperatures during the day.

morning   **noon**   **late afternoon**

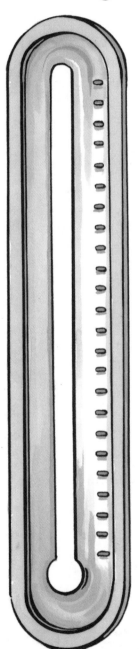 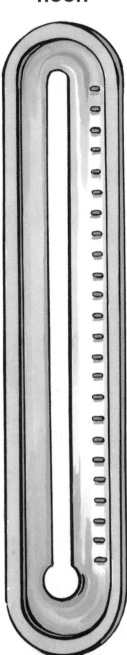 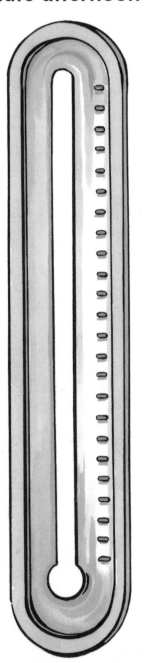

Focus | Children use colour to show relative temperatures on thermometers.

Copyright © 2005 Pearson Education Canada Inc. Not to be copied.   **Unit 3, Lesson 8:** Show What You Know   **73**

Name: _____   Date: _____

My Journal

Which clock do you prefer: or [6:15] ?
Tell about your thinking.

_____

_____

_____

_____

How can you tell if the temperature is rising?

_____

_____

_____

_____

Tell what you learned about counting coins.

_____

_____

_____

_____

HOME CONNECTION
Invite your child to share what he or she liked about this unit.

Focus | Children reflect on what they have learned about time, temperature, and money.

The Skating Day

The class looked outside at the ice and the snow.
"If this storm doesn't stop soon, I don't think we'll go."
Then Cam's grandma came. "All the roads are okay.
We'll still take the bus to go skating today."

It was so frosty cold when they got on the bus,
they climbed on it quickly without any fuss.
Miss Chu called out names, as they passed her in line.
She didn't want any children left behind.

The volunteers helped them lace their skates on tight.
It was lots of hard work to tie them just right.
"It's freezing!" said Mishi. Her teeth loudly chattered.
"I love skating!" cried Cam. To him, nothing else mattered.

Once on the ice, it was easy to spot
the children who skated and those who did not.
Grandma flew by; she was spinning and turning.
She encouraged the children: "Once I was just learning."

Then Grandma stopped skating. "My scarf! Where's it gone?
I'm sure that this morning I put the scarf on."
They looked on the rink, but the scarf wasn't there.
They looked by the benches; they looked everywhere.

"Who'd want my scarf? That is puzzling to me."
Cam tugged Grandma's coat. Mishi said, "It was me.
I felt freezing cold, so I wrapped myself in it.
I thought I would borrow it just for a minute."

Grandma smiled. "I'm so glad that the scarf is not gone.
It's my favourite for winter, but you keep it on.
Let's skate a while longer, before leaving the rink.
Then, to warm up, we'll have hot chocolate to drink!"

About the Story

The story was read in class to prepare for a Mathematics Investigation activity. Children completed addition and subtraction activities, worked with number combinations, and counted in a variety of ways. They also identified and made their own patterns and used coins to represent different money amounts.

Talk about It Together

- What happens to Grandma's scarf while the class is at the ice arena?
- How do you think the parent volunteers might be using math during the field trip?
- How did Mishi feel when she realized she had the scarf?
- What kind of person is Grandma? Do you know someone like her? How is that person the same? Different?

At the Library

Ask your local librarian about other good books to share about patterning, numbers, and measuring time, temperature, and money.

How Can We Arrange 24 Children?

There are 24 children and 3 parts to the rink.
Show one way to arrange the children into 3 groups.

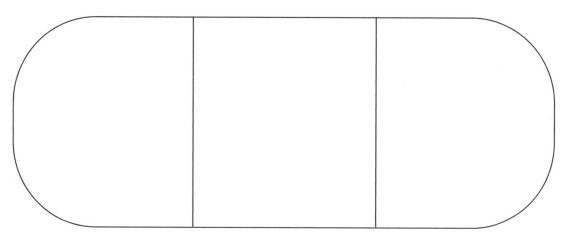

Show a different way to arrange 24 children in 3 groups.

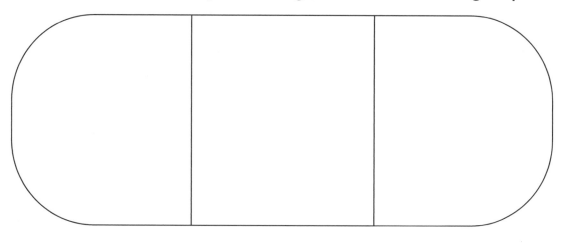

Grandma's Scarf

What is the pattern on Grandma's scarf?
Show the pattern, using 3 repeats.

Draw a circle around the pattern core.

Make your own pattern for a scarf.
Draw your pattern.

Describe your pattern.

_____

_____

How Many Skates?

Show 2 ways to count the skates.

How many skates are there altogether? _____
Are there more hockey skates or figure skates?

_____

How many more? _____
Show how you know. Use pictures, numbers, or words.

How many skates are needed for the children in your class?
Show your thinking. Use pictures, numbers, or words.

Buy Cam a Snack!

Grandma has $1.00 in coins.
Draw the coins that she could have in her purse.

Grandma can spend up to
$1.00 on Cam's snack.

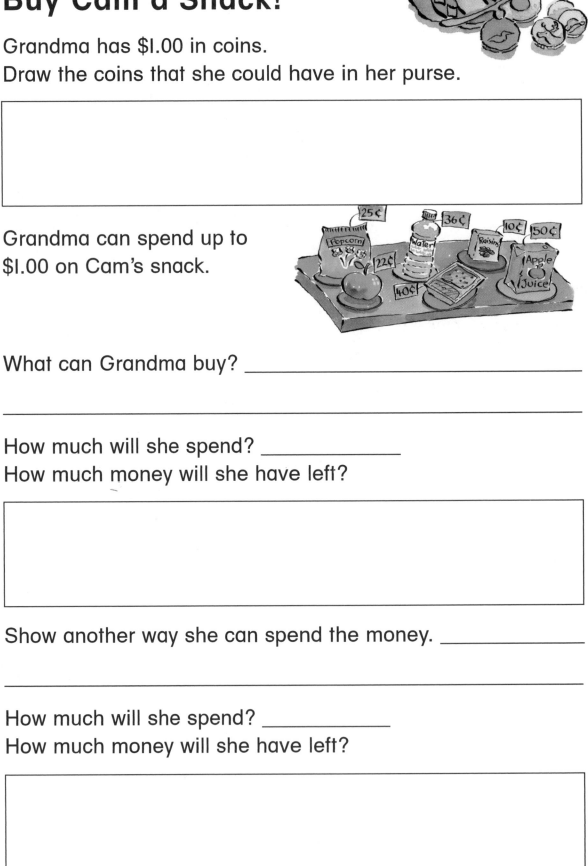

What can Grandma buy? _____

_____

How much will she spend? _____
How much money will she have left?

Show another way she can spend the money. _____

_____

How much will she spend? _____
How much money will she have left?

86

Floating Bubbles

Challenge a friend to see
whose bubble stays
in the air longer.
Blow at the same time
and begin counting slowly.

Do you think
you will get to 10? 20?
Will you get as high
as 50?

Food for Thought

Fill in the blanks with something that makes sense.

I could eat 100 _____

but not 100 _____

I could lift 100 _____

but not 100 _____

I would like to have 100 _____

but not 100 _____

Make up some more sentences of your own.

_____

_____

The next 4 pages fold in half to make an 8-page booklet.

Fold

Math at Home

Hop aboard
the Math Express.
Where will we go today?
To a land of math mystery
where we can learn and play.

We'll see patterns and calendars
and numbers galore.
We'll see money and measuring,
so much fun in store!

Pattern Search

At home, look for patterns made with

- lines
- squares
- different shapes
- numbers

Can you find other types of patterns?

Elevating Elevators

Imagine you are in a really tall building. You leave the doctor's office and get on the elevator at the 21st floor. You need to go down 14 floors to get to the cafeteria.

Which button will you push?

Suppose you got on at the 17th floor and went up 18 floors. On which floor will you get off?

Make up some elevator problems of your own.

Crazy Cookies

The cookie-making machine at a local factory has gone wild! Each time a cookie pops out, its shape changes. Check the first 3 cookies that came out.

1st   **2nd**   **3rd**

What will the 5th cookie look like?
What will the 7th cookie look like?
Which one would you like to eat?

Find Your Page!

Find a book with a few hundred pages in it. Get a friend to call out a page number that could be in the book.

Open the book as close as you can to that page number. Estimate how far off you were, then give your friend a turn.

What was the closest you got?
Would it be easier if the book had 50 pages? Why?

Plenty of Time

Think of things you do every day and decide whether they take less than a minute, about a minute, or more than a minute to finish.

Use a chart to record your predictions.

Less than a minute	About a minute	More than a minute

All done? Put your list on the refrigerator. Next time you do one of the activities, check to see if your prediction was right!

When Were You Born?

Find out when each person in your family was born.

Whose birthday comes first in the year?

Whose birthday comes after yours?

Is there a month with more than one birthday?

Make a list or a picture to show what you found out.

Dirty Laundry

Max ran through a big puddle and then shook off the mud right beside the clean laundry.

Estimate how many mud spots landed on the sheet. (Think groups of 10s and the estimating will be a breeze!)

How Much Farther?

Next time you are in the car or on the bus, be on the lookout for place signs that show distance away in kilometres.

Which place is nearest?

Which is farthest?

How much farther?

Create other problems with these distances.

MILLWOOD 3 km
BLUEVALE 10 km
ACTON 14 km

Ten-Frames for Secret Numbers

Can you think of a different game to play with the same materials?

Game

Secret Numbers

You'll need:
- 5 sets of number cards from 0 to 9, shuffled and placed face down
- ten-frames
- 20 counters

On your turn:
- Draw 2 cards and place them **face up** in front of you, like this.
- Draw 2 more cards and place them **face down** in front of a friend.

2nd card	1st card
10s place	1s place

Your friend has a choice:
- Trade a card for one of yours.
- Leave the cards alone.

Once you trade or keep cards, look at the numbers.
- Flip over the face-down cards and then read both numbers.
- The player with the greater number puts a counter on his or her ten-frame.
- The cards go into the discard pile and the other player draws the next cards.

The first player to fill a ten-frame wins!

Exploring Addition and Subtraction

Focus | Children create number stories about a scene from a country fair.

91

Dear Family,

In this unit, your child will be developing strategies for adding and subtracting two-digit numbers.

The Learning Goals for this unit are to

- Add three addends (for example, 5 + 6 + 7 = 18).
- Use 10 to help when adding and subtracting.
- Add multiples of 10 to one- and two-digit numbers.
- Develop and use different strategies to add and subtract pairs of two-digit numbers.
- Look for patterns in digits when adding or subtracting.

You can help your child achieve these goals by doing the Home Connection activities suggested at the bottom of selected pages.

Number Stories from the Fair

Write two number sentences about each picture.

_____ + _____ = _____

_____ − _____ = _____

_____ + _____ = _____

_____ − _____ = _____

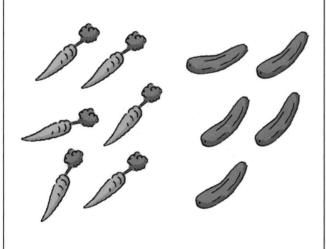

_____ + _____ = _____

_____ − _____ = _____

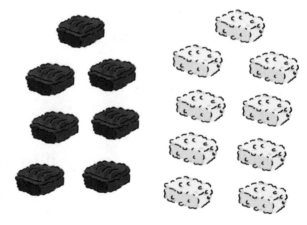

_____ + _____ = _____

_____ − _____ = _____

Adding Rows and Columns

Look at the tables of numbers.
Add the numbers in each row.
Add the numbers in each column.

column

2	3	5
8	2	7
6	1	8

row → ___
row → ___
row → ___

___ ___ ___

column

5	7	5
7	3	2
4	6	8

row → ___
row → ___
row → ___

___ ___ ___

When did the "find 10" strategy help you?

_____

_____

When did the "near doubles" strategy help you?

_____

_____

What other strategies did you use?

_____

_____

_____

HOME CONNECTION
Use the first three digits of phone numbers to write and solve addition sentences with your child. For example, if a phone number begins with 747, the sentence is 7 + 4 + 7 = 18.

FOCUS | Children use different strategies to add three numbers.

Adding Animals

Animals We Saw at the Fair

Animal Groups	Animals	Number
Goats		28
		17
		19
Cattle		31
		30
Horses		56
		13
Poultry		48

Choose an animal group. Make an addition story.
Show how you solved the problem. Use pictures, numbers, or words.

Focus | Children add two-digit numbers using their own strategies.

Adding Your Way

The fair has a contest for the longest zucchini.
Omar grew a zucchini 43 cm long.
The winning zucchini is 11 cm longer.
How long is the winning zucchini?

Tell how you solved the problem.
Use pictures, numbers, or words.

There are 38 gasoline tractors in the tractor show.
There are 16 steam-powered tractors.
How many tractors are there altogether?

Tell how you solved the problem.
Use pictures, numbers, or words.

Focus | Children add two-digit numbers using their own strategies.

The Ferris Wheel

There are 28 people on the Ferris wheel.
45 more are waiting in line.
How many people are at the ride altogether?

Tell how you solved the problem.
Use pictures, numbers, or words.

Make up your own addition story.

_____ + _____ = _____

Find the Way Home

Help the cow find its way home to the barn.
Find each sum. Circle the pairs of numbers that add to 73.
Join the circled numbers to draw the path home.

$64 + 9 =$ _____

$56 + 9 =$ _____

$57 + 17 =$ _____

$50 + 23 =$ _____

$27 + 18 =$ _____

$40 + 23 =$ _____

$42 + 31 =$ _____

$37 + 25 =$ _____

$32 + 41 =$ _____

$43 + 30 =$ _____

$62 + 10 =$ _____

$28 + 45 =$ _____

Look at the pairs of numbers that add to 73.
Which sums were easy for you to find? Why?

_____

_____

HOME CONNECTION
Have your child explain how he or she found the answer to $64 + 9$. Then,
work together to think of three more pairs of numbers that add to 73.

Focus | Children add two-digit numbers
using their own strategies.

Give Me 10

Choose a number between 1 and 9.

Write it as the starting number in the table below.

Add 10 to your number.

Write the sum in the first row of the table.

Use the sum as the starting number in the next row.

Add 10 each time. Keep going until you have filled all the rows.

Number	Add 10	Sum
	+ 10	
	+ 10	
	+ 10	
	+ 10	
	+ 10	
	+ 10	

What stays the same when you add 10 to a number?

_____

What changes?

_____

Predict what will happen if you add 10 three more times.

_____

Check your prediction. Write the number sentences.

____ + ____ = ____   ____ + ____ = ____   ____ + ____ = ____

Focus | Children add 10 to a number several times and record the pattern.

Dime Addition

Add dimes to each bank.
Write the total amount of money in each bank.

_____¢ in all

_____¢ in all

_____¢ in all

_____¢ in all

_____¢ in all

_____¢ in all

HOME CONNECTION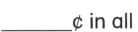
Place 6 pennies in a row. Ask your child: "How many pennies are there?" Add a dime to the row and ask: "How much money is there altogether?" Continue adding dimes until you reach 96¢.

Focus | Children add dimes to find money amounts up to 99¢.

Adding 10s

Complete the addition sentences.

26 + 10 = _____   81 + 10 = _____   18 + 70 = _____

9 + 90 = _____   31 + 40 = _____   23 + 60 = _____

$$
\begin{array}{r} 48 \\ + 30 \\ \hline \end{array}
\qquad
\begin{array}{r} 17 \\ + 60 \\ \hline \end{array}
\qquad
\begin{array}{r} 10 \\ + 40 \\ \hline \end{array}
\qquad
\begin{array}{r} 15 \\ + 50 \\ \hline \end{array}
$$

$$
\begin{array}{r} 29 \\ + 30 \\ \hline \end{array}
\qquad
\begin{array}{r} 11 \\ + 50 \\ \hline \end{array}
\qquad
\begin{array}{r} 16 \\ + 80 \\ \hline \end{array}
\qquad
\begin{array}{r} 12 \\ + 70 \\ \hline \end{array}
$$

8 + 10 = _____   13 + 30 = _____   14 + 10 = _____

Suppose you add 20 to a number.
Predict how the number will change.

_____

_____

How can you check your prediction?

_____

_____

Focus | Children add groups of 10 to one- and two-digit numbers.

How Many Cakes?

The fair has an Ugly Cake contest for children.
There are 48 cakes entered in the contest.
Then, Mr. Melnik's class enters 6 more cakes.
How many cakes are there altogether?

Describe how you would find the answer.
Use pictures, numbers, or words.

```



```

Now, find these sums.

8	18	28	38
+ 6	+ 6	+ 6	+ 6

Look at all the sums on this page. How are they alike?

_____

_____

Predict the answer to 58 + 6. Add to check.

_____

Addition Patterns

Find the sums.

9	19	29	39	49	59
+ 5	+ 5	+ 5	+ 5	+ 5	+ 5

All the answers _____.

What is the answer to 79 + 5? _____
How do you know?

_____

Find the sums.

4	4	4	4	4
+ 18	+ 28	+ 38	+ 48	+ 58

All the answers _____.

What is the missing number in 4 + ☐ = 92? _____
How do you know?

_____

HOME CONNECTION

Ask your child to explain how he or she found the missing number in 4 + ☐ = 92.

Focus | Children add one- and two-digit numbers and look for patterns in the sums.

How Many More?

There are more cats than rabbits entered in the show. How many more?
Show how you solved the problem. Use pictures, numbers, or words.

_____ – _____ = _____

Focus | Children look for information in a story and subtract two-digit numbers using their own strategies.

HOME CONNECTION
Work with your child to write a different subtraction-story problem using the information about the pet show.

To the Fair

There are 38 Grade 2 children going to a country fair.
Some Grade 1 children are also going.
Altogether, 62 children will go.
How many Grade 1 children are going?

Tell how you solved the problem.
Use pictures, numbers, or words.

Make up your own subtraction story.

_____ − _____ = _____

Focus | Children solve subtraction stories using their own strategies.

Left in the Line

There are 39 people lined up
to ride the Rocket Blaster.
The ride ends and 16 people from the line get on.
How many are left waiting in the line?

Tell how you solved the problem.
Use pictures, numbers, or words.

Make up your own subtraction story.

_____ − _____ = _____

A Number Code

Ryan and Ayesha made up a code
that uses numbers in place of letters.
Here are some of the letters and numbers they use.

C	E	G	H	I	N	O	P	R	S	W
3	5	7	8	9	14	15	16	18	19	20

Ryan wrote these numbers: 16   9   7

What animal name do they spell? _____

Ayesha wrote subtraction sentences as clues for her favourite
farm animal. Solve each clue.

34 – 15 = _____   The letter is _____.

67 – 59 = _____   The letter is _____.

49 – 44 = _____   The letter is _____.

61 – 56 = _____   The letter is _____.

78 – 62 = _____   The letter is _____.

What is Ayesha's favourite farm animal? _____

What animal name do these differences spell? _____

22 – 19 = _____   28 – 13 = _____   55 – 35 = _____

How Many Now?

There are 56 children in the Fun Run.
9 children stop for a drink.
How many students are still running?

Describe how you would find the answer.
Use pictures, numbers, or words.

Now, find these differences.

$$
\begin{array}{r} 16 \\ -\ 9 \\ \hline \end{array}
\qquad
\begin{array}{r} 26 \\ -\ 9 \\ \hline \end{array}
\qquad
\begin{array}{r} 36 \\ -\ 9 \\ \hline \end{array}
\qquad
\begin{array}{r} 46 \\ -\ 9 \\ \hline \end{array}
$$

Look at all the differences on this page. How are they alike?

_____

_____

Predict the answer to 76 − 9. Then subtract to check.

_____

Focus | Children use their own strategies to subtract one-digit numbers from two-digit numbers. Then, they find related differences and look for patterns.

Subtraction Patterns

Find the differences.

8	18	28	38	48	58
− 6	− 6	− 6	− 6	− 6	− 6
_____	_____	_____	_____	_____	_____

All the answers _____.

What is the answer to 78 − 6? _____
How do you know?

_____

Find the differences.

14	24	34	44	54
− 8	− 8	− 8	− 8	− 8
_____	_____	_____	_____	_____

All the answers _____.

What is the missing number in ☐ − 8 = 76? _____
How do you know?

_____

HOME CONNECTION

Ask your child to explain how he or she found the missing number in ☐ − 8 = 76.

Focus | Children subtract one-digit numbers from two-digit numbers and look for patterns in the differences.

Name: _____   Date: _____

Feed the Pigs

Helping to feed the pigs is one of Hannah's farm chores. She puts ears of corn in buckets for them.

Bucket 1 has 37 ears. Bucket 2 has 41 ears. There are 95 ears of corn altogether. How many ears of corn are in Bucket 3?

Show how to solve the problem. Use pictures, numbers, or words.

Focus | Children use a problem-solving strategy of their choice to solve a story problem.

How Many Ears of Corn?

A farmer has 95 ears of corn.
The mother pig eats 46 ears.
One young pig eats 27 ears.
How many ears of corn does the other young pig eat?

Show how to solve the problem.
Use pictures, numbers, or words.

HOME CONNECTION

Ask your child: "How can you check if your solution to the problem is correct?"

Focus | Children use a problem-solving strategy of their choice to solve a story problem.

The Answer Is 53

Make 2 addition stories that have a sum of 53.
Use pictures, numbers, or words.
Write an addition sentence for each story.

_____ + _____ = 53

_____ + _____ = 53

FOCUS | Children demonstrate their understanding of addition by creating their own addition stories.

112   Unit 4, Lesson 8: Show What You Know   Copyright © 2005 Pearson Education Canada Inc. Not to be copied.

Subtraction Stories for 53

Make 2 subtraction stories that begin with 53.
Use pictures, numbers, or words.
Write a subtraction sentence for each story.

53 – _____ = _____

53 – _____ = _____

Name: _____ Date: _____

My Journal

Tell what you learned about adding large numbers.
Use pictures, numbers, or words.

┌───┐
│   │
│   │
│   │
│   │
│   │
│   │
│   │
└───┘

Tell what you learned about subtracting large numbers.
Use pictures, numbers, or words.

┌───┐
│   │
│   │
│   │
│   │
│   │
│   │
└───┘

HOME CONNECTION
With your child, talk about situations at home when
you would need to use addition or subtraction.

Focus | Children reflect on and record what they learned about different ways to add and subtract large numbers.

Unit 4, Lesson 8: Show What You Know

Data Management and Probability

Always, Sometimes, Never!

Race to ten
is a game we will play.

I wonder which colour
will **likely** win today?

Tiles in a bag—
shake them all about.

Put your hand in
and pull one out.

Will it **sometimes** be yellow
or will it **never** be blue?

Will it **always** be red?
What is your clue?

Always, sometimes, never
are the words you will use.

Look at your tile and
record what you choose.

The first colour to ten
is the winner you see!

Always, sometimes, never—
what will it be?

Focus | Children predict and describe the outcome of a game.

Dear Family,

In this unit, your child will
be learning about making graphs
and probability—how likely an event is.

The Learning Goals for this unit are to

- Collect, organize, describe, and label data on graphs.
  For example, your child's class may tally the number of
  days there is rain, snow, sun, or clouds during a month
  and graph the results.
- Read graphs and ask questions about the data they
  gathered.
- Talk about probability in day-to-day situations using
  words such as *always*, *sometimes*, or *never* to describe
  events. For example: It sometimes snows in spring.
- Use mathematical language such as *likely*, *unlikely,* and
  probably when playing games. For example: I think the
  spinner will likely land on red.

You can help your child reach these goals by doing the Home
Connection activities suggested at the bottom of selected pages.

Race to Ten

Which colour will win the race?
Circle your answer. **red** or **blue**

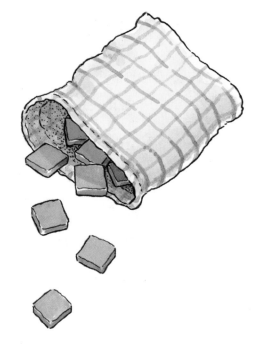

Game Directions

Place 10 red tiles
and 5 blue tiles in a bag.

Take a tile from the bag.

Colour one square on the path.

Put the tile back in the bag. Repeat.

Race to ten!

red

blue

Which colour won? _____

How do you know? _____

Why do you think that happened? _____

_____

Focus | Children pull a tile from a bag, colour a square on the path, replace the tile, and repeat until one row is filled.

Unit 5, Launch: Data Management and Probability **117**

Build the Snow Pictures

Choose a spinner.
Circle your choice.

Print **A** on one part of the spinner.
Print **B** on the other part.

You will spin 10 times.
Which snowperson will **likely** be taller, **A** or **B**?

Spin 10 times.
Which snowperson was taller? Why do you think that happened?

Choose a different spinner.
Play again.

Which snowperson do you think will be taller, **A** or **B**?

Which snowperson was taller? Why do you think that happened?

Focus | Children build cut-out pictures from *LM 6: Build the Snow Pictures* according to spinner outcomes.

118 Unit 5, Lesson 1: Probability   Copyright © 2005 Pearson Education Canada Inc. Not to be copied.

The Tortoise and the Hare Race

Choose a spinner.
Circle your choice. ○ ○ ○ ○

Print **T** for **tortoise** on one part.
Print **H** for **hare** on the other part.

Who do you think will win?
Spin the spinner. Colour one square for each spin.

Tell what happened. _____

_____

Choose a different spinner. ○ ○ ○ ○
Play again.

Who do you think will win?
Spin the spinner. Colour one square for each spin.

Tell what happened. _____

_____

Focus | Children choose and label a spinner from *LM 5: Spinners*, predict the winner, and record results.

Likely or Unlikely?

Look at each picture. Circle likely or unlikely.

likely unlikely	likely unlikely
likely unlikely	likely unlikely
likely unlikely	likely unlikely

Make Your Own   likely   unlikely

HOME CONNECTION
Make a likely/unlikely book with your child. Fold paper in half and have your child draw likely events on one side and unlikely events on the other.

Focus | Children look at the picture and decide whether it is likely or unlikely. They create their own likely and unlikely pictures.

Make a Game

Which spinner will you use in your game?
Circle it.

What do the players try to do in your game?

What will **probably** happen if each player spins 10 times?
Tell about your thinking.

Play your game with a partner. Tell what happened.

Focus | Children design a game with a spinner, predict the results, and play the game, explaining the results.

Copyright © 2005 Pearson Education Canada Inc. Not to be copied.   Unit 5, Lesson 2: Strategies Tool Kit   121

Name: _____   Date: _____

Change the Spinner

Play your game again.
Choose a different spinner.

Circle it.   ⊘

Predict what will **probably** happen
when you play your game using this spinner.

Tell about your thinking.

Play your game with a partner. Tell what happened.

Make a Bar Graph

Toss a two-coloured counter 15 times.

Predict which colour will come up most often.
Circle **red** or **yellow**.

Make a tally chart.

Colours	Tally
red	
yellow	

Make a bar graph to show
the results.

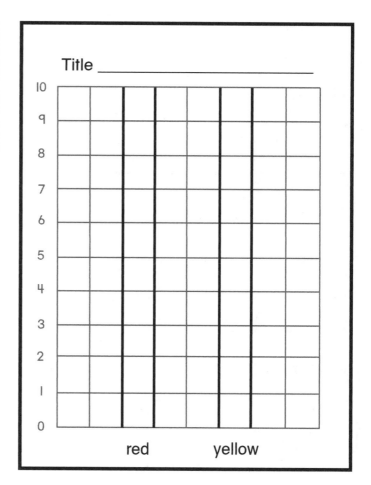

Title _____

red   yellow

What did you find out?

_____

_____

_____

Focus | Children predict the results of tossing a two-colour counter, and record results in a tally and on a bar graph.

Name: _____   Date: _____

Tally and Count

I counted all the chairs in our classroom.

My tally is _____.

There are _____ chairs.

I counted all the _____ in our classroom.

My tally is _____.

There are _____.

I counted all the _____ in our classroom.

My tally is _____.

There are _____.

How did tallying help you with your counting?

```



```

Focus | Children make a tally of different items in the classroom.

124   Unit 5, Lesson 3: Making a Bar Graph

A Fishy Graph!

How many fish have stripes?
How many fish have spots?
How many fish have whiskers?

Make a tally chart.

Type	Tally
stripes	
spots	
whiskers	

Make a bar graph to show the results.

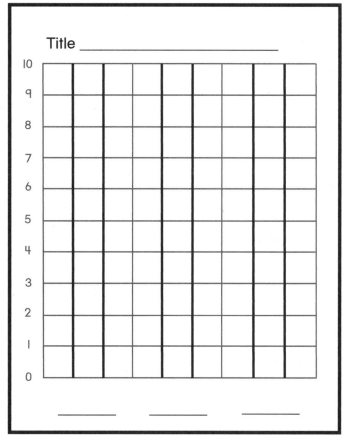

Title _____

Write a true statement about the fish.

HOME CONNECTION
Practise making a tally at home with your child, for example, by tallying all the spoons in the drawer, or the books on a shelf.

Conduct a Survey

My Question _____

Ask 10 friends. Make a tally chart.

Choices	Tally

What did you find out?

_____

_____

What do you think would happen if you asked
the same question in a Grade 5 class?

HOME CONNECTION

Have your child ask a question of the family, such as: "Do you
like the snow?" Have your child ask up to 10 family members
or friends, then make a tally and bar graph of the results.

Focus | Children conduct a survey and record
the results in a graph.

Name: _____ Date: _____

What Is Your Favourite?

My Question

Do you like _____, _____, or _____ the best?

Ask 10 friends. Make a tally chart.

Types of Food	Tally

Make a bar graph to show
the results.

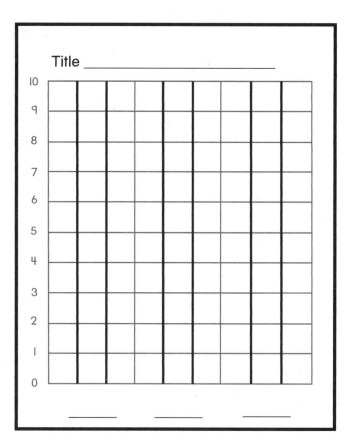

Title _____

What did you find out? _____

_____

What did you learn about conducting a survey?

_____

_____

Name: _____ Date: _____

Dinosaur Graph

Ari's Toy Dinosaurs

What does this graph tell you?

_____

What questions can you ask about it?

_____

_____

_____

What did you learn from this graph?

I learned that _____

_____

I learned that _____

_____

At the Park

Do you like to go to the park?

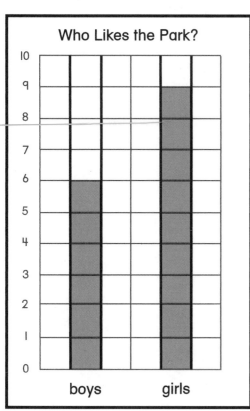

What is the graph about? How do you know?

_____

_____

One question I have about the information is

_____

_____

Focus | Children read and interpret a bar graph. They ask a question about the information shown.

Copyright © 2005 Pearson Education Canada Inc. Not to be copied.   Unit 5, Lesson 5: Interpreting a Graph   129

Greater or Less?

Before you play the game, predict.
Will most sums be greater or less than your number?

Circle **greater** or **less**.

Explain your thinking in pictures, numbers, or words.

```
┌─────────────────────────────────────────────┐
│                                             │
│                                             │
│                                             │
│                                             │
│                                             │
│                                             │
│                                             │
└─────────────────────────────────────────────┘
```

Make a graph on the next page to show your results.

Did your results match your predictions?   **yes**   **no**

Make 2 true statements about your results.

_____

_____

If you play the game again, what do you think will happen?
Use pictures, numbers, or words to explain why.

```
┌─────────────────────────────────────────────┐
│                                             │
│                                             │
│                                             │
│                                             │
│                                             │
│                                             │
└─────────────────────────────────────────────┘
```

Focus | Children choose a number between 9 and 15, then turn over two cards and record whether the sum is greater or less than their number. They continue until they have used all of the cards; then they graph the results.

Name: _____ Date: _____

Our Results

Play "Greater or Less?"
Make a tally chart.

Sums	Tally
sums less than _____	
sums of _____	
sums greater than _____	

Make a bar graph to show the results.

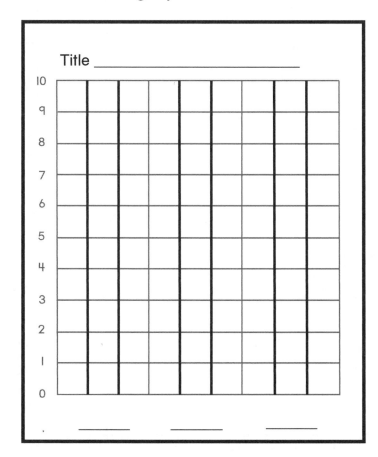

Title _____

10
9
8
7
6
5
4
3
2
1
0

_____ _____ _____

HOME CONNECTION

Use game board number cubes to play a game with your child. Even or odd? Roll the number cubes, and find the sum. If it's odd, one player gets a point. If it's even, the other player gets a point.

Focus | Children create a tally and bar graph to display the results of their "Greater or Less?" game.

Name: _____ Date: _____

My Journal

Tell what you have learned about spinners and how they work.

_____

_____

_____

_____

_____

What are graphs used for?

_____

_____

_____

_____

_____

Focus | Children use pictures, numbers, or words to reflect on what they have learned about data management and probability.

3-D Geometry

Focus | Children identify and describe geometric solids.

133

Dear Family,

In this unit, your child will be
learning about 3-D objects, such as
cubes, spheres, cylinders, prisms, and pyramids.

The Learning Goals for this unit are to

- Describe, compare, and sort 3-D solids according to
  their attributes, such as whether they roll or stack,
  have curved or flat surfaces.
- Use 3-D solids in constructions.
- Make skeletons to represent 3-D solids.
- Use language such as *pyramid, prism, face,* and *edge* to
  describe solids.

You can help your child achieve these
goals by doing the Home Connection
activities suggested at the
bottom of selected pages.

Name: _____   Date: _____

Our Construction

We made a _____ .

We used these solids.

Solid	How Many Solids?

We used _____ solids in all.

Focus | Children construct a 3-D model of a building and record the number of each type of solid they used.

Unit 6, Launch: 3-D Geometry   **135**

A Sorting Rule

Make a sorting rule.

My sorting rule is _____

_____.

Circle the objects that fit your rule.

What other way can you sort the solids?

Write another sorting rule. _____

_____

Put an ✗ on the objects that fit your new rule.

Name: _____ Date: _____

Solids That Are Alike

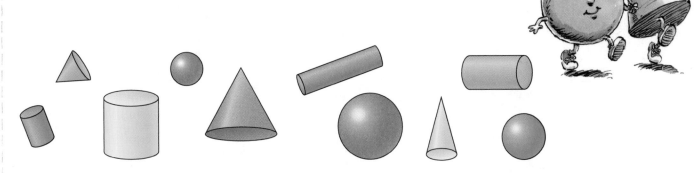

All these solids have _____.

They can all _____.

All these solids have _____.

They can all _____.

In which group would you put a cube? Explain.

_____

_____

Focus | Children identify shared attributes for groups of objects.

Same and Different

What solid did your teacher give you?

Circle it. Underline its name in the chart.

Choose another solid.

Circle it. Underline its name in the chart.

Count the faces, edges, and vertices on your solids. Fill in the chart.

	Rectangular Prism Triangular Prism Pyramid	Cone Cylinder Sphere Cube
Number of Faces		
Number of Edges		
Number of Vertices		

How are the two solids the same? _____

_____

How are they different? _____

_____

Focus | Children record the numbers of faces, edges, and vertices in each of two solids. Then they compare the solids according to their attributes.

How Many Faces?

A ▲ has 5 faces.
Draw the faces you traced.

How many ■ faces? _____   How many ▲ faces? _____

A ▭ has 6 faces.
Draw the faces you traced.

How many ■ faces? _____   How many ▭ faces? _____

How are the two solids alike?

_____

How are they different?

_____

Focus | Children trace around the faces of two solids and identify the number of each type of face. Then they compare the solids.

Unit 6, Lesson 2: Comparing Solids   **139**

Use the Clues

Use geometric solids like these.

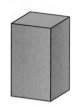

rectangular prism

triangular prism

pyramid

cone

cylinder

sphere

Follow the clues to fill in the chart.

Clue	Solid
greatest number of faces	
only one vertex (point)	
two circular faces	
five vertices (points)	
two triangle faces	
no vertices (points)	

Focus | Children use clues about faces, edges, and vertices to identify 3-D solids.

The Construction Challenge

Circle the structure you are going to build.

| the widest structure |
| the strongest structure |
| a lookout tower |

| a castle with at least one pyramid |
| a structure with a ramp |

| a structure with only prisms |
| the tallest structure |
| a structure with two cylinders |

| a structure with exactly eight solids |
| a structure that uses one of each of the solids |

Build your structure.

Tell how you made your structure.
Use pictures, numbers, or words.

Unit 6, Lesson 3: Building with 3-D Solids

Name: _____ Date: _____

The Best House

Which house would you choose?
Circle your choice.

Explain your choice. _____

_____

Show how to build the house using solids.
Use pictures, numbers, or words.

Focus | Children choose their favourite house from three pictures. They explain their choice and tell how to build it with solids.

Modelling Clay Solids

Which picture did you choose?
Circle it.

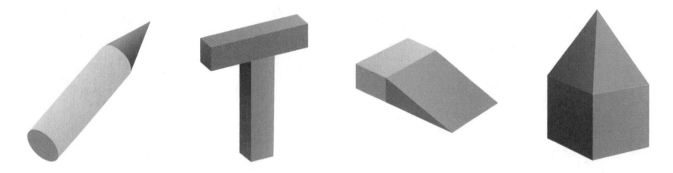

How did you use solids to make your model?
Use pictures, numbers, or words to explain.

Making Solids

Use modelling clay to make one of these solids.

Show how you made your solid.
Give tips that would help a friend make it.

Use pictures, numbers, or words.

Which solids are best for building? Why? _____

_____

Focus | Children construct a solid from modelling clay and explain how they did it.

144   **Unit 6, Lesson 4:** Build a Model   Copyright © 2005 Pearson Education Canada Inc. Not to be copied.

Name: _____   Date: _____

See the Solids

How many pyramids and prisms do you see?

	Number of Rectangular Prisms	Number of Triangular Prisms	Number of Pyramids

HOME CONNECTION
Have your child work with modelling clay to practise making various solids. Ask how many faces, edges, and corners your child sees for each solid.

Skeletons

cube   triangular   rectangular   rectangular
              prism   prism   pyramid

Match each skeleton to a solid.

This is the skeleton of a _____.

How do you know? _____

_____

This is the skeleton of a _____.

How do you know? _____

_____

This is the skeleton of a _____.

How do you know? _____

_____

Focus | Children name the solid represented by each skeleton model and explain their thinking.

Build a Skeleton

Use straws and clay to build a skeleton.

I built a skeleton for a _____.

How many did you use?

I used _____ long straws.

I used _____ short straws.

I used _____ pieces of clay.

Write two tips for building the skeleton.

1. _____

_____

_____

2. _____

_____

_____

A _____ cannot be built as a skeleton.

Why? _____

HOME CONNECTION
Ask your child: "What steps did you use to build the skeleton?"

How Many for a Cube?

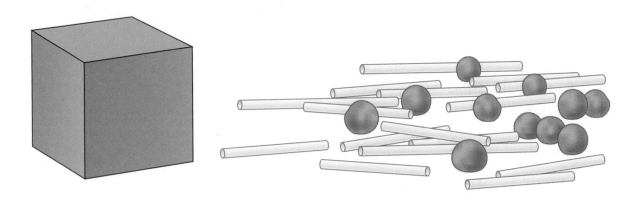

Plan to make a skeleton for a cube.

How many straws and pieces of clay will you need?

Fill in the chart.

Solid	Number of Straws	Number of Pieces of Clay
cube		

Tell how you solved the problem. Use pictures, numbers, or words.

How Many for a Prism or a Pyramid?

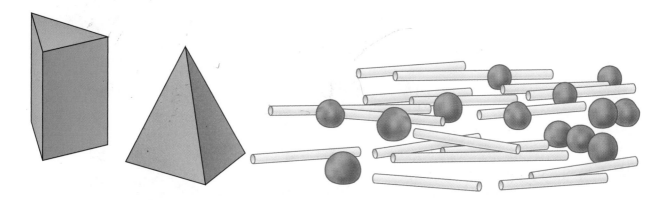

Choose one solid.

How many straws and pieces of clay will you need
to make its skeleton?

Fill in the chart.

My Solid	Number of Long Straws	Number of Short Straws	Number of Pieces of Clay

Tell how you solved the problem. Use pictures, numbers, or words.

Focus | Children determine how many straws and pieces of clay as joiners they need to build the skeleton of a solid. Then they record how they solved the problem.

Pack the Spaceship

The astronauts are taking these supplies on their trip.

The astronauts need to pack the supplies in three bins.

Choose a sorting rule for each bin.
Decide which objects to put in.

Circle the objects that go in Bin 1.
The objects go together because they all have _____.

Write an **✗** on objects that go in Bin 2.
The objects go together because they all have _____.

Place a **✔** on objects that go in Bin 3.
The objects go together because they all have _____.

Focus | Children sort objects into three bins and explain their reasoning.

Build a Spaceship

Which solids will you need
to build this spaceship?

Make them from
modelling clay.
Put them together
to match the picture.

Think of another structure
you can build.
Use solids to make it.

How many solids did you use? _____

Solid	Number of Solids in the Spaceship	Number of Solids in My Structure
cube		
rectangular prism		
triangular prism		
pyramid		
cone		
cylinder		
sphere		

Focus | Children build two structures and record how many of each type of solid they used in their structures.

Unit 6, Lesson 7: Show What You Know **151**

Name: _____   Date: _____

My Journal

Tell about the spaceship and the structure you built.
Use pictures, numbers, or words.

What did you learn about solids?
Use pictures, numbers, or words to show your thinking.

Focus | Children tell about their spaceship and the structure they built. Then they relate what they learned about 3-D solids in the unit.

HOME CONNECTION
Invite your child to help put away groceries. Talk about why you place particular packages or supplies together.

Addition and Subtraction to 100

Focus | Children identify the different ways to show 67.

Dear Family,

In this unit, your child will be learning about addition and subtraction of two-digit numbers, and developing an algorithm for finding and recording solutions.

The Learning Goals for this unit are to

- Model addition and subtraction of two-digit numbers on a place-value mat.
- Use the standard algorithm as one way to add and subtract two-digit numbers.
- Pose and solve number problems requiring addition or subtraction.
- Use a calculator to solve addition or subtraction problems with numbers over 50.

You can help your child achieve these goals by doing the Home Connection activities suggested at the bottom of selected pages.

Take a Number

Choose a number from the team line-up. Circle it.

53 15 27 76 88

Show your number at least 2 ways.

[ ]

Choose another number. Put a square around it.
Compare this number to your first choice.
Use pictures, numbers, or words.

[ ]

Find 2 numbers that come between your numbers.

Focus | Children represent a number in different ways, and compare numbers.

Name: _____ Date: _____

Same Number, Different Ways

Choose a two-digit number.
Model the number on a place-value mat.
Draw a picture of your place-value mat.

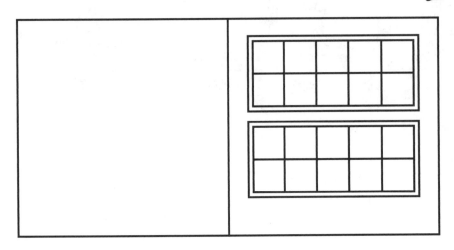

_____ 10s _____ 1s _____ in all

Use the same two-digit number.
Model the number a different way on a place-value mat.
Draw a picture of your place-value mat.

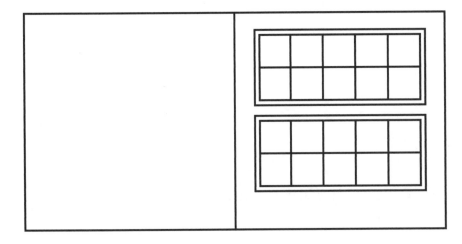

_____ 10s _____ 1s _____ in all

Make a Match

Look for matching place-value mats.
Join each pair by drawing a line.

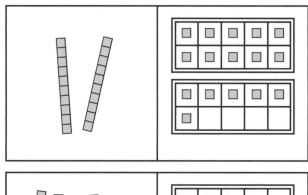

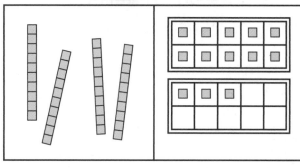

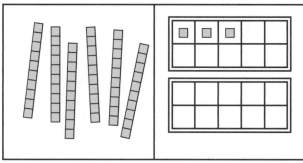

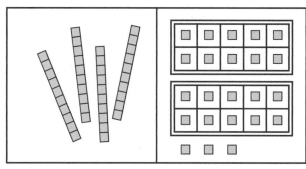

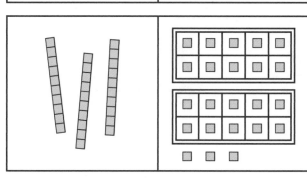

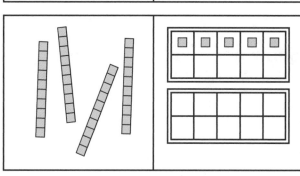

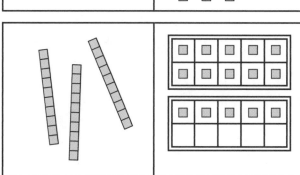

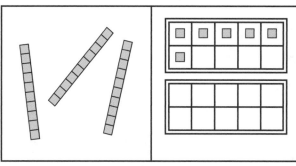

HOME CONNECTION

Make paper clip or elastic band chains to represent 26 (2 chains of 10 and 1 chain of 6). Ask your child to write the number in 3 ways.

FOCUS | Children identify equivalent representations of numbers displayed on place-value mats.

On with the Show!

The Grade 2 children put on a show.
They gave out 28 blue tickets.
They gave out 37 yellow tickets.
How many tickets did they give out in all?

Use materials. Solve the story problem.
Show your answer.

Show another way to solve the problem.

Focus | Children use base ten concepts and materials to solve an addition problem.

158 Unit 7, Lesson 2: Adding Two-Digit Numbers Copyright © 2005 Pearson Education Canada Inc. Not to be copied.

At the Show

The Grade 2 show was held in the gym.
The children helped to bring 55 chairs to the gym.
They still need 25 chairs.
How many chairs do
they need altogether?

Use materials.
Solve the story problem.
Show your answer.

The Grade 2 children served refreshments after the show.
Jordan poured cups of lemonade.
Kiya handed out 24 cups of lemonade.
18 cups of lemonade
were still on the table.
How many cups of lemonade
did Jordan pour?

Solve the story problem.
Show how you solved it.

Make your own story problem. Solve it.

_____

_____

_____

_____

Focus | Children create and solve story problems involving addition, using place-value concepts.

Unit 7, Lesson 2: Adding Two-Digit Numbers

Clean-Up Time

After the show, the Grade 2 children helped clean up.
There were

    28 red balloons
    28 blue balloons
    36 white streamers
    18 blue streamers
    37 clean cups
    43 dirty cups

Make a story problem about the Grade 2 clean-up.
Solve your problem. Show your work in pictures, numbers, or words.

Make another story problem.
Change books with a partner. Solve your partner's problem.

Partner's Name
_____

Focus | Children create and solve story problems involving addition, using place-value concepts.

It All Adds Up

Which addition sentences do you think will need a trade?
Circle them.

Use materials and a place-value mat.
Answer the addition problems you circled.

22 + 23 = _____   67 + 26 = _____   53 + 20 = _____

9 + 40 = _____   34 + 57 = _____   19 + 49 = _____

Write the problems that do not need a trade.
Solve them.

_____

_____

How did you know which problems needed a trade?

_____

_____

HOME CONNECTION
Provide straws or toothpicks that can be clustered into 10s with a twist tie. Ask your child to model one of the addition problems from this page.

How Many Shells?

Laura collects shells.
She has 36 shells.
She finds 28 more on the beach.
How many shells does Laura have?

Build the addition story on a place-value mat.
Record it here.
Then record the addition using numbers.

How are the two ways of finding the answer the same?

_____

_____

How are they different?

_____

_____

Focus | Children use materials to solve an addition problem using the standard algorithm, and record their work.

162 Unit 7, Lesson 3: Recording Addition with the Standard Algorithm   Copyright © 2005 Pearson Education Canada Inc. Not to be copied.

In the Schoolyard

There are 37 children in the schoolyard.
The bus arrives with 25 more children.
How many children are in the schoolyard now?

Build the addition story on a place-value mat.
Record it here.
Then record the addition using numbers.

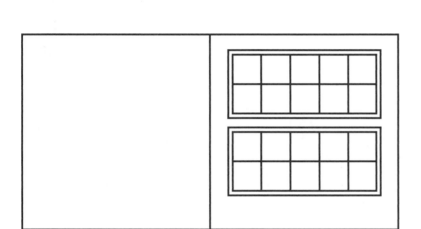

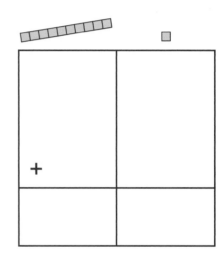

Build these addition stories.
Record the answers on the charts.

7	4
+ 1	7

2	8
+ 2	3

3	9
+ 2	0

3	1
+ 1	9

Addition Stories

Build these addition stories on a place-value mat.
Record the answers on this page.

$$\begin{array}{r} 28 \\ +11 \\ \hline \end{array} \qquad \begin{array}{r} 34 \\ +23 \\ \hline \end{array} \qquad \begin{array}{r} 44 \\ +18 \\ \hline \end{array} \qquad \begin{array}{r} 89 \\ +16 \\ \hline \end{array}$$

$$\begin{array}{r} 26 \\ +34 \\ \hline \end{array} \qquad \begin{array}{r} 52 \\ +18 \\ \hline \end{array} \qquad \begin{array}{r} 46 \\ +35 \\ \hline \end{array} \qquad \begin{array}{r} 19 \\ +65 \\ \hline \end{array}$$

What other way can you solve 52 + 18?
Show your way using pictures, numbers, or words.

Choose an addition story you can solve a different way.
Show your way using pictures, numbers, or words.

Name: _____   Date: _____

Read All about It!

Leo has 65 comic books.
He gave 28 of them to Lizzie.
How many comic books does Leo have now?

Use materials. Solve the story problem.
Show how you solved it.

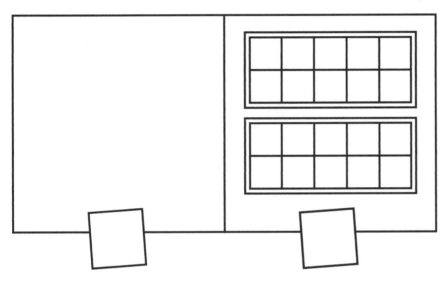

Can you think of another way to solve the problem?
Show it here.

Unit 7, Lesson 4: Subtracting Two-Digit Numbers

Bookworm Bonanza

Leo and Lizzie joined a comic book club.
The book club has 96 comic books to share.
The first week, Lizzie borrowed 18 of them.
How many were left?

Use materials.
Solve the story problem.
Show how you solved it.

Another week, Leo brought in 24 comic books.
Lizzie brought in 17.
Who brought in more?
How many more?

Solve the story problem.
Show how you solved it.

Make your own story problem. Solve it.

Focus | Children create and solve story problems using place-value concepts.

166   Unit 7, Lesson 4: Subtracting Two-Digit Numbers   Copyright © 2005 Pearson Education Canada Inc. Not to be copied.

The Bookworm List

Our Top Five

Superheroes Beware - 48 pages
Once Again – – – – – 76 pages
Back in Time – – – 32 pages
Take-Off ! – – 56 pages
Countdown – 96 pages

Make a story problem about the book list.

Solve your problem.

Show your work in pictures, words, or numbers.

Make another story problem.

Change books with a partner.

Solve your partner's problem.

Partner's Name

_____

Focus | Children create and solve story problems using place-value concepts. Then they solve a partner's problem.

Copyright © 2005 Pearson Education Canada Inc. Not to be copied.   Unit 7, Lesson 4: Subtracting Two-Digit Numbers   **167**

What's the Difference?

Which subtraction stories do you think will need a trade?
Circle them.

Use a place-value mat to solve the questions you circled.

69 – 39 = _____   53 – 14 = _____   65 – 26 = _____

33 – 18 = _____   64 – 57 = _____   93 – 30 = _____

Write the problems that do not need a trade.
Solve them.

HOME CONNECTION
Have your child tell about one number sentence that required trading. Ask your child: "How did the place-value mat help you find the answer?"

How Many Fossils?

Dr. Rockhound's 53 fossils
are made up of shells and rocks.
Naveen counted 27 shells.
How many rocks does Dr. Rockhound have?

Build the subtraction story on a place-value mat.
Record it here.
Then record the subtraction using numbers.

How are your picture and your
number recording alike?

_____

How are they different?

_____

Focus | Children use materials to solve a subtraction problem using the standard algorithm, and record their work.

At the Museum

There are 75 children at the museum.

36 children visit the fossils.

The others visit the cave exhibit.

How many children are at the cave exhibit?

Build the subtraction story on a place-value mat.

Record it here.

Then record the subtraction using numbers.

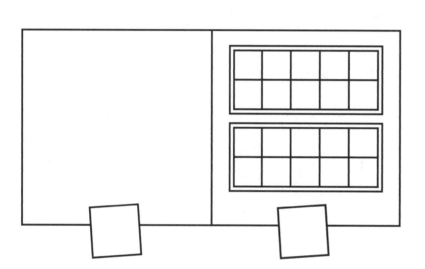
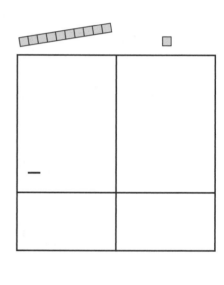

Find the answers.

Use your base 10 materials for each.

7	8
− 4	5

3	2
− 1	3

5	5
− 2	8

6	7
− 2	8

Subtraction Stories

Build these subtraction stories.
Use Snap Cubes and a place-value mat.
Record the answers on this page.

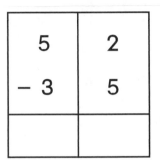

5	2
− 3	5

8	7
− 3	7

6	5
− 3	8

8	6
− 5	9

1	7
−	9

7	0
− 1	9

8	9
− 1	8

8	0
− 7	5

Show another way to find 70 − 19. Use pictures, numbers, or words.

Choose a subtraction story you can solve a different way.
Show it using pictures, numbers, or words.

Focus | Children use materials to build subtraction stories using the standard algorithm. They then solve subtraction stories in different ways.

HOME CONNECTION
Have your child describe two ways to solve 80 − 75, with and without materials.

What's in Pat's Pockets?

Pat has 4 coins in her pocket.
What could the coins be?
How much money might Pat have?

Show how you solved the problem.
Use pictures, numbers, or words.

Focus | Children choose 4 coins and find their combined value.

Name the Coins

There is 83¢ in Joe's piggy bank.
He empties his bank and counts the coins.
What could the coins be?

Show how to solve the problem.
Use pictures, numbers, or words.

HOME CONNECTION
Take turns with your child creating money riddles. Model a money
amount without showing it, then give the total and the number of coins.

Focus | Children choose a strategy
to solve a problem.

At the Book Fair

Jenn unpacked the 55 nature books.
She unpacked the 28 puzzle books.
How many books did Jenn
unpack altogether?

Use materials. Show your answer.

Rae put price stickers on all 78 picture books.
Jenn put price stickers on all 28 puzzle books.
How many more books did Rae stick with stickers?

Use materials. Show your answer.

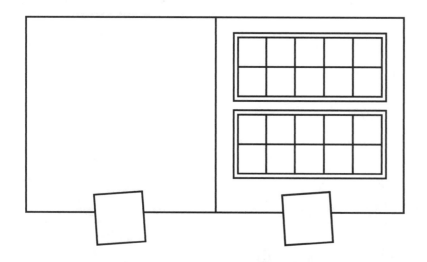

Focus | Children solve story problems involving addition and subtraction, using place-value concepts by modelling on a place-value chart. They record solutions pictorially.

Books on Order

	Monday	**Tuesday**	**Wednesday**
number of books ordered	26	43	79

Make an addition story about the books ordered.
Show your addition story on a place-value mat.
Record the addition using numbers.

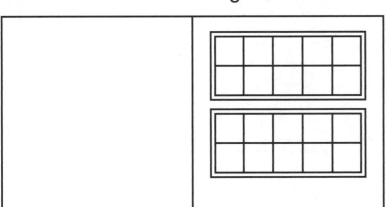

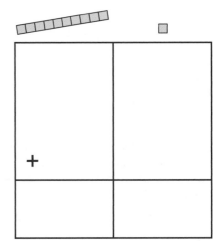

Make a subtraction story about the books ordered.
Show your subtraction story on a place-value mat.
Record the subtraction using numbers.

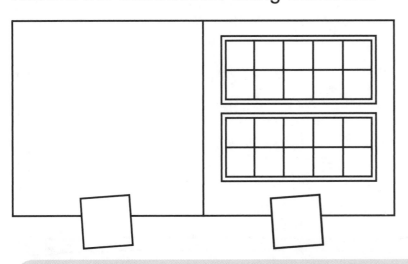

Focus | Children create and solve story problems related to addition and subtraction of two-digit numbers. They record solutions pictorially and symbolically.

Name: _____ Date: _____

My Journal

Tell what you learned about adding.
Use pictures, numbers, or words.

Tell what you learned about subtracting.
Use pictures, numbers, or words.

Planning "Spring Fling"

Cam ran through the door. "I'm home!" he called out.
The dog came to see what the noise was about.
"We're having a 'Spring Fling'—a school celebration.
I've brought you your own special 'Fling' invitation!"

The next day Grade 2 started planning and thinking.
"What snacks will we serve? What will people be drinking?
How will we decorate? Where will we sit?"
Grandma thought, "Maybe I'll help out a bit."

Miss Chu said, "We'll greet people when they arrive.
If everyone comes, there will be 75!
We'll hand out the programs we make at the door.
We'll keep extras here in this box on the floor."

"We'll need a large space that can fit us all in.
We can hang up spring posters to show in the gym!
How will we set up the tables and chairs?
I hope there's enough. I hope we have spares!"

They brainstormed together for more than an hour.
"We should practise a song for the Primary choir.
Reciting a spring poem would be special, too.
Are there any more things the Grade 2 class might do?"

They posted a graph that they studied a lot.
Cam asked, "Is there anything that we forgot?"
They grew more excited as every day passed.
Then finally the day came. They all sighed, "At last!"

They'd planned very well. They were all well prepared.
But they ran out of programs, so everyone shared.
The families grew quiet. The choir came in.
"Welcome," Cam said. "Let the 'Spring Fling' begin!"

About the Story

The story was read in class to prepare for a Mathematics Investigation activity. Children asked questions about and made predictions from a bar graph. They added and subtracted two-digit numbers, recognized a growing pattern, and counted by 10s.

Talk about It Together

- Have you ever planned an event like "Spring Fling" before? What are the kinds of things you had to think about? How did knowing math help you?
- Is there anything you would have done differently than the Grade 2 class?
- How were the students feeling as they planned the event?
- What is your favourite part of the story?

At the Library

Ask your local librarian about other good books to share about numbers, collecting and analyzing information, and geometric shapes.

What Will We Do?

Look at the graph.
Write 3 things you can learn from the graph.

1. _____

2. _____

3. _____

The Grade I class made a graph about what they would like to do for the "Spring Fling."

Would the graph be the same or different than the Grade 2 graph?

In what ways would the graph be the same or different?
Use pictures, numbers, or words to explain your thinking.

```
Cass
Tom
Helena
Ming
Jovan      Sherrill
Devon      Abasi        Lisa
Marnie     Chin         Zane
Sami       Pedro        Jack
Nic        Lana         Nina
Cam        Kali         Ron
                        Anwar
                        Ana
                        Blake
Song       Poem         Dance
```

_____
_____
_____
_____
_____
_____
_____
_____

Think of something different the Grade 2 class could make a graph about to help them in their planning.

_____

How Many Are Coming?

All the primary classes invited their families.
Some moms and dads are coming.
Some grandparents and friends are coming, too.

Look at the tally chart.
Count the tally marks for each class.
Write the number of guests in the last column.

Class	Tally of Guests	How Many Are Coming?
Kindergarten	⊞ ⊞ ⊞ ⊞ ⊞	_____ guests
Grade 1	⊞ ⊞ ⊞	_____ guests
Grade 2	⊞ ⊞ ⊞ ⊞ ⊞ ⊞ ⊞	_____ guests

How many guests are coming altogether? _____
Show how you solved the problem.

Show two ways to arrange the guests' chairs in groups.

I made my groups this way because _____

_____.

I made my groups this way because _____

_____.

Write a subtraction problem about the guests and show how to solve it.

How Many Cans?

The school is having a food drive.
Families bring 92 cans of food.
The children stack the cans in pyramids of 10s.

How many pyramids of 10s can they build? _____

How many cans will be left over? _____
Show your thinking in pictures, numbers, or words.

Describe the pattern you see in the pyramid.

_____

Suppose the children add another row to one pyramid.

How many cans of food would be in this new row? _____

How many cans of food would be in the new pyramid? _____

Suppose the children continue to add rows to one pyramid.

How many rows would the largest pyramid have? _____
Show your thinking in pictures, numbers, or words.

Crazy Container Tally

Which types of food packaging
are most popular in your home?

Cans? Boxes? Jars?
Plastic containers?

Get ready to investigate
by printing each category
on a piece of paper.

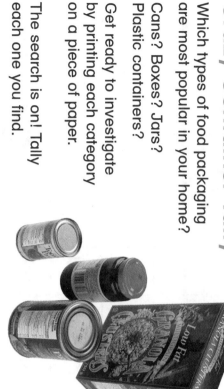

The search is on! Tally
each one you find.

When you are done, use all the information
to create a "Food Package" bar graph.

Tell someone about what you found.
Which was most popular? Least popular?
Were you surprised?

Name the Missing Buttons

Seven buttons are in a bag.

One button was taken out and then put back in.
Here are eight draws.

What colours do you think the last two buttons were?

The next 4 pages fold in half to make an 8-page booklet.

Fold

Math at Home

Let's whip up
some math today.
The recipe sounds great!
Just add eight
scoops of numbers,
centimetres, and a date.
Toss in a bunch of solids,
perhaps a coin or two.
I'll stir in a juicy survey.
Mmm. "Math Stew"!

What's the Difference? Game Board

10	11	12	13	14	15	16	17	18	19
20	21	22	23	24	25	26	27	28	29
30	31	32	33	34	35	36	37	38	39
40	41	42	43	44	45	46	47	48	49
50	51	52	53	54	55	56	57	58	59
60	61	62	63	64	65	66	67	68	69
70	71	72	73	74	75	76	77	78	79
80	81	82	83	84	85	86	87	88	89
90	91	92	93	94	95	96	97	98	99

Mystery Solids

Put 5 or 6 small 3-D solids
into a bag you cannot see through.
Reach in and take 1 of the solids.
Challenge a friend to guess
your solid by asking yes
or no questions.
Only 10 questions are allowed.

If your friend guesses
the solid, show it.
Then give the bag to your
friend and play again.
You guess this time!

Adding Stars

How many stars altogether?

Sean said, "8 and 2 more is 10. And 4 more is 14."

Amy said, "I know 8 and 8 is 16.
If I take 2 away, it's 14."

Who's right? Why do you think so?

What's the Difference?

You'll need:

- game board (page 7)
- two sets of number cards, 1 to 9 and one 0 card, in a bag you cannot see through
- 6 counters
- paper
- pencil

On your turn:

- Draw 2 cards and use them to make a two-digit number. Put a counter on the matching number on the game board.

- Draw 2 more cards and make another two-digit number. Use a counter to cover this number on the game board.

- Find the difference between the 2 numbers by counting how many 10s and 1s they are away from each other. Use a counter to cover this number.

If the numbers **33** and **57** were covered, you would say, "57 is two 10s and four is away from 33. The difference is 24."

If the difference is

- more than 25, you get a point.
- a number with a ☆, you get a point.
- a number with a ♥, your friend gets a point.

After tallying the points, place the cards back in the bag and remove your counters from the board.

Take turns until someone gets 10 points.

Copyright © 2005 Pearson Education Canada Inc. Not to be copied.

At the Fair

How many different ways could you win this carnival game?

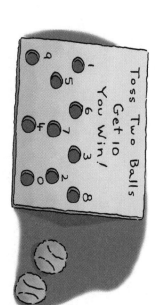

Toss Two Balls
Get 10
You Win!

1 5 9
6 4 7
3 2
0 8

Snack Time

You want to buy an apple and you have this much money.

75¢

Do you have enough?
If not, how much more
do you need?

Out at the Park

Next time you are at the park, look around at the play equipment.
What shapes do you see?
Do you notice any patterns?

Addition Chart for Four in a Row

+	0	1	2	3	4	5	6	7	8	9
0	0	1	2	3	4	5	6	7	8	9
1	1	2	3	4	5	6	7	8	9	10
2	2	3	4	5	6	7	8	9	10	11
3	3	4	5	6	7	8	9	10	11	12
4	4	5	6	7	8	9	10	11	12	13
5	5	6	7	8	9	10	11	12	13	14
6	6	7	8	9	10	11	12	13	14	15
7	7	8	9	10	11	12	13	14	15	16
8	8	9	10	11	12	13	14	15	16	17
9	9	10	11	12	13	14	15	16	17	18

Four in a Row Game

You'll need:
- addition chart (page 5)
- number cards 1 to 9
- two kinds of small counters (beans, buttons)

Put the addition chart and cards between both players. Cards should be face down.

Each player gets a pile of counters.

On your turn:
- Draw 2 cards and say 2 addition sentences that match the numbers.
- Put one of your counters on either of the matching sums on the chart.

So, if you drew **8** and **6**, you would put your counter on the sum of 8 + 6 or 6 + 8.

- Put the cards back down and move them around.

Take turns until someone gets 4 counters in a row.

Remember to try to block the other player!

What patterns do you see?

Linear Measurement, Area, and Perimeter

FOCUS | Children look for ways people are measuring to prepare classroom decorations.

Name: _____ Date: _____

Dear Family,

In this unit, your child will be learning about measurement. Your child will develop an understanding of linear measurement and area.

The Learning Goals for this unit are to

- Estimate, measure, and compare lengths using non-standard units, such as paper clips, straws, or Snap Cube trains.
- Develop an understanding of the need for standard units and measure lengths using the centimetre and the metre.
- Estimate, measure, and compare perimeter (distance around) using non-standard and standard units.
- Investigate area measurement by using non-standard units, such as cards, to cover a surface.
- Solve everyday problems about measurement.

You can help your child achieve these goals by doing the Home Connection activities suggested at the bottom of selected pages.

My Banner Measurements

Draw your desk. Where will you put your banner?
Tell how you measured it.
Use pictures, numbers, or words.

Measure Three Ways

Choose two objects to measure.

I chose _____ and _____.

Measure each object 3 times, using ☐ ☐ ☐ .
Estimate before each measure.
Complete each table.

My first object is _____.

Measuring Unit	Estimate	Measure

My second object is _____.

Measuring Unit	Estimate	Measure

Compare your measurements. What do you notice?

_____

_____

Focus | Children estimate and measure the lengths of two objects using a variety of non-standard units and compare results.

Estimate, Measure, and Compare

Choose four books.

What will you use to measure their heights?

Circle one.

Complete the table.

Book Title	Height

Put the books on the shelf in order of their heights.
What other ways can you order the books?

_____

_____

_____

What other ways can you measure the books?

_____

_____

_____

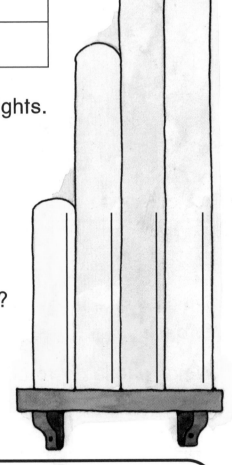

HOME CONNECTION

Choose something in your home that you and your child can order together by length (for example, clothes in a closet, towels on a rod).

Focus | Children measure and compare the heights of four books.

Real-Life Beetles

The pictures match the size of each beetle in real life.
About how long is each one?

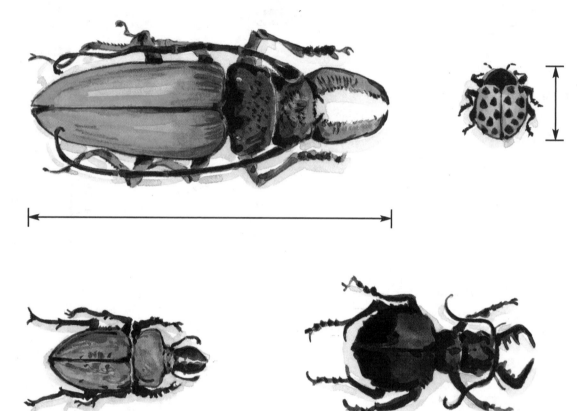

Which beetle is the longest?
Circle it.

Which beetle do you think is the widest?
Measure to check. Were you right?

_____

_____

Focus | Children measure and compare the lengths of spiders using a centimetre ruler.

Calling all 10s

Find four objects that you think are each about 10 cm long.

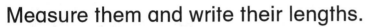

Measure them and write their lengths.

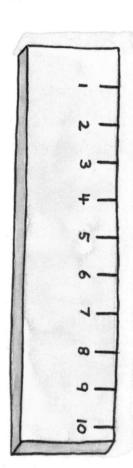

Object	Length
	_____ cm
	_____ cm
	_____ cm
	_____ cm

What helped you to find objects about 10 cm long?

_____

_____

Focus | Children estimate and then use a ruler to measure four objects, each about 10 cm long.

About How Long Is a Metre?

Choose three objects in the classroom.
Look for

• one object that is shorter than I m long
• one object that is longer than I m long
• one object that is about I m long

Complete the table.
Name each object and estimate its length.
Then, measure it.

	My Object	**Estimate**	**Measure**
shorter than I m			
longer than I m			
about I m			

Focus	Children choose objects that are longer than, shorter than, and about, I m long. They estimate and measure the length of each object.

Which Unit?

What would you use to measure these objects?
Circle centimetre or metre.

centimetre   metre

centimetre   metre

centimetre   metre

centimetre   metre

centimetre   metre

Choose one of your answers. Tell about your thinking.

_____

_____

HOME CONNECTION

Have your child estimate and measure the heights of family members using metres, centimetres, or both.

Focus | Children decide whether they would use the centimetre or the metre to measure each of the pictured objects.

Name: _____ Date: _____

Poster Space

Measure the wall spaces in your classroom.

Record each measure in the table.

Wall Space	Width
1	
2	
3	

Put the widths in order from least to greatest.

_____   _____   _____

Focus | Children order the widths of specified wall spaces in their classroom to determine which one will best fit a poster.

Measure and Graph

Find 4 objects to measure.

books   **magazines**

shoes

your choice

pencil cases

Complete the table.

My Objects	Measure
	_____ cm
	_____ cm
	_____ cm
	_____ cm

Make a graph to show
your measurements.

Write your measurements from shortest to longest.

_____   _____   _____   _____

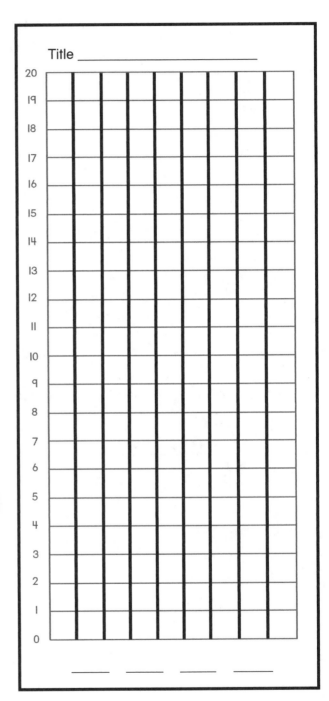

Title _____

20
19
18
17
16
15
14
13
12
11
10
9
8
7
6
5
4
3
2
1
0

_____ _____ _____ _____

Focus | Children measure 4 objects, then graph and order their results.

Unit 8, Lesson 5: Compare and Order Lengths   **203**

Distance Around

Draw pictures of the figures you measured.
Inside each figure, write the distance around.

[blank box]

Use pictures, numbers, or words to tell how you found the
distances around.

[blank box]

Focus | Children find the perimeter of several large figures on the classroom floor.

HOME CONNECTION
Choose a photograph with your child and have your child
find the distance around it. Then, measure and cut paper
to make a decorative frame.

Frame It

Find the distance around each picture.
Measure 3 times.

distance around _____ cm

distance around _____ cm

distance around _____ cm

Which picture has the shortest distance around? Circle it.

Focus | Children measure distances around pictures.

Unit 8, Lesson 6: Distance Around

Cover It

You need to tile the dollhouse floor.
Choose one figure to cover the floor.
This is your unit. Circle it.

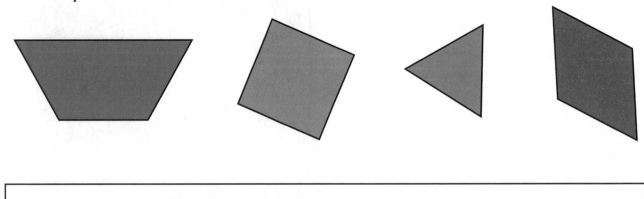

How many units do you think you will need? _____
Cover the floor.
How many units did you need? _____

About How Many?

About how many blocks will cover each surface?

Object	Estimate	Measure
the top of your desk	about _____	
a box lid	about _____	
my object	about _____	

Suppose you measured again with another unit.
What do you think would happen?

_____

_____

Focus | Children use materials to measure the area of surfaces in the classroom.

HOME CONNECTION
Have your child find the area of a tabletop or counter using index cards. Use the measurement to create a paper cover for the table or to make placemats.

How Many Ways?

Make different arrangements
with 5 Snap Cubes.
How many arrangements can you find?
Show all your arrangements.

Triangle Figures

Make different arrangements with 5 Pattern Block triangles.
Record your arrangements.

HOME CONNECTION
Use a word game that has letter tiles, and with your child, create the names of family members. For each name, look for different ways to arrange the number of tiles used.

Name: _____   Date: _____

Our Obstacle Course

Measure the distance around your course in 2 ways.
Use pictures, numbers, or words to show how you measured.

First way

```


```

What is the distance around? _____

Second way

```


```

What is the distance around? _____

Measure the area your course covers. Use any way you like.
Show what you did in pictures, numbers, or words.

How we measured the course

```


```

How big is the course? _____

Focus | Children measure distance around and the area of an obstacle course. They choose their own measuring units and instruments.

Making the Obstacle Course

Estimate how long the course is altogether.

_____

Measure to find out. _____

_____

What part did you make more than I m long? _____

How long is it? _____

_____

What part is more than 5 cm high? _____

How high is it? _____

_____

What part is more than I0 cm wide? _____

How wide is it? _____

_____

How far is it from the start to the first obstacle? _____

_____

What is the tallest part to climb? _____

How tall is it? _____

How do you know it is the tallest? _____

Name: _____   Date: _____

My Journal

Tell what you have learned about measuring lengths.

> _____
>
> _____
>
> _____
>
> _____
>
> _____

Tell what you have learned about covering surfaces.

> _____
>
> _____
>
> _____
>
> _____

HOME CONNECTION
Ask your child: "Why do you think learning about measurement is important?"

Focus | Children tell what they have learned about linear measurement, area, and perimeter.

2-D Geometry and Patterning

Focus | Children talk about the picture and identify the 2-D figures they recognize.

Dear Family,

In this unit, your child will be learning more about 2-D geometric figures, symmetry, and patterns.

The Learning Goals for this unit are to

- Describe, sort, and compare 2-D figures according to the number of sides and number of vertices (corners).
- Recognize figures that have symmetry, or matching parts.
- Make patterns using materials and pictures of 2-D figures.
- Use positional language to describe movement and location: *slide to the left, turn, above, beside.*

You can help your child achieve these goals by doing the Home Connection activities suggested at the bottom of selected pages.

Name: _____ Date: _____

My Picture

Use ▪ ● ▭ and ▲ .

Make a picture.

```




```

Tell about the figures you used.

_____

_____

Focus | Children create pictures using 2-D figures and describe their work.

Name: _____   Date: _____

Where Am I?

Look in your classroom.
Find an example of each figure.
Tell where you saw each one.

rectangle   square   triangle   hexagon   octagon   pentagon   circle

What I Saw	It Looks Like	Where I Saw It
book cover	a rectangle	inside my desk

Look for It!

Colour a figure with 3 sides  red .

Colour a figure with 4 vertices blue .

Colour a pentagon green .

Colour a hexagon Purple .

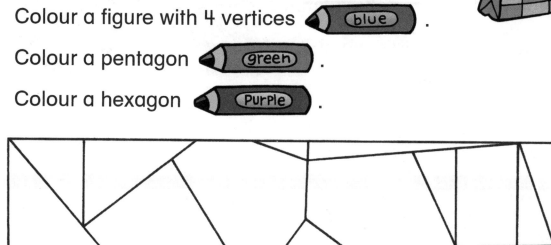

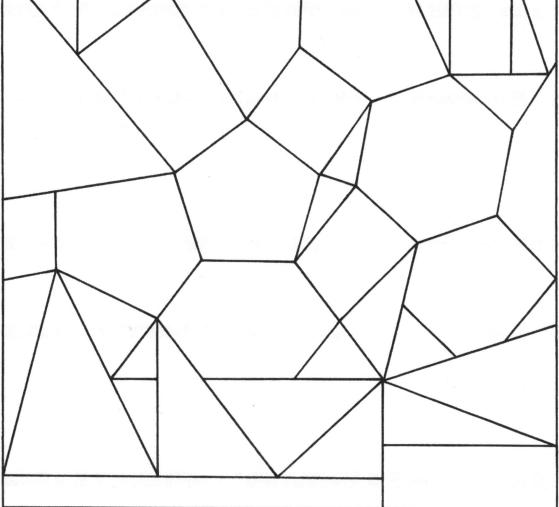

HOME CONNECTION

With your child, look for examples in your home of some of the figures your child has been learning about, such as circles, triangles, rectangles, pentagons, hexagons, and octagons. Which do you find most often?

Focus | Children look for and colour figures in a drawing.

Make a Match!

Choose a figure.
Find another child with a matching figure.

How do you know your figures match?
Use pictures, numbers, or words.

HOME CONNECTION

Cut out several pairs of matching triangles (same size and shape). Scramble them. Pick one triangle. Ask your child to find its match. After all are paired, ask: "What strategies did you use to match the triangles?"

Focus | Children each choose a figure from a set and then look for a matching figure. They describe how they know that the figures match.

Alike and Different

Circle 2 figures.

I chose the _____ and the _____.

One way they are alike is _____

_____.

Another way they are alike is _____

_____.

One way they are different is _____

_____.

Focus | Children select 2 figures and describe how they are alike and different.

What Is in the Bag?

There are 3 figures in a bag.
The total number of sides is 13.

What could the figures be?
Show your strategy.

What other figures could they be?

Focus | Children choose a strategy to find 3 figures with a total number of 13 sides.

Another 3 in the Bag

There are 3 figures in a bag.
The total number of vertices is 12.

What could the figures be?
Show your strategy.

What other figures could they be?

Focus | Children choose a strategy to find 3 figures with a total number of 12 vertices.

HOME CONNECTION
With your child, go on a vertex (corner) hunt in your home. Look for a geometric figure and count its corners. Ask: "How can we find another figure with the same number of corners?"

Matching Parts

Circle the figures that have matching parts.
Draw the fold lines.

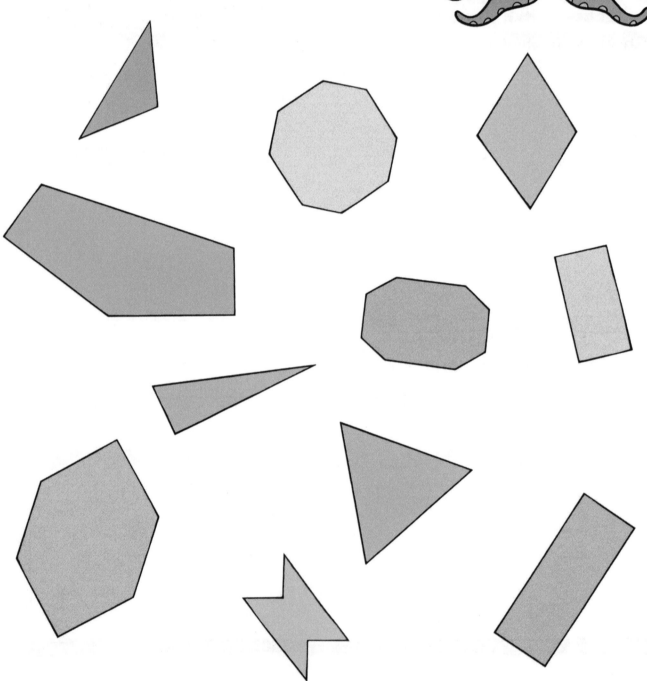

Name: ______________________________ Date: ______________________________

Symmetry Cutouts

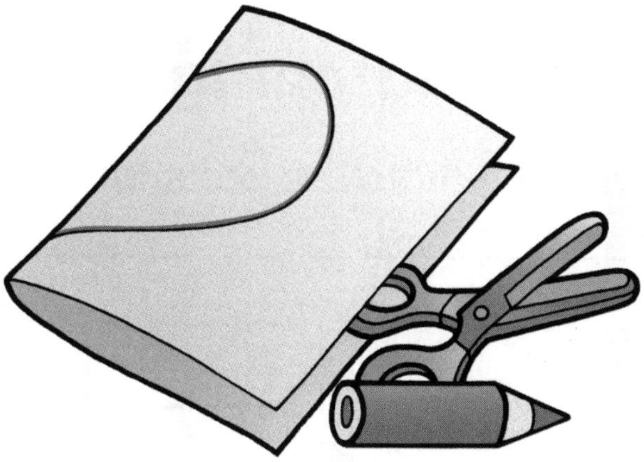

Fold a piece of paper in half.
Draw an outline for half a figure.
Begin on the fold line.
End on the fold line.

Keep your paper folded.
Cut on the outline. But do not cut on the fold line!
Open the paper and paste the figure below.

What happened when you opened up the paper?

____

Focus | Children create a symmetrical figure by folding paper and cutting. Then they answer a related question.

Draw Matching Parts

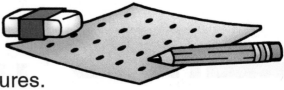

Draw the matching parts to finish the pictures.

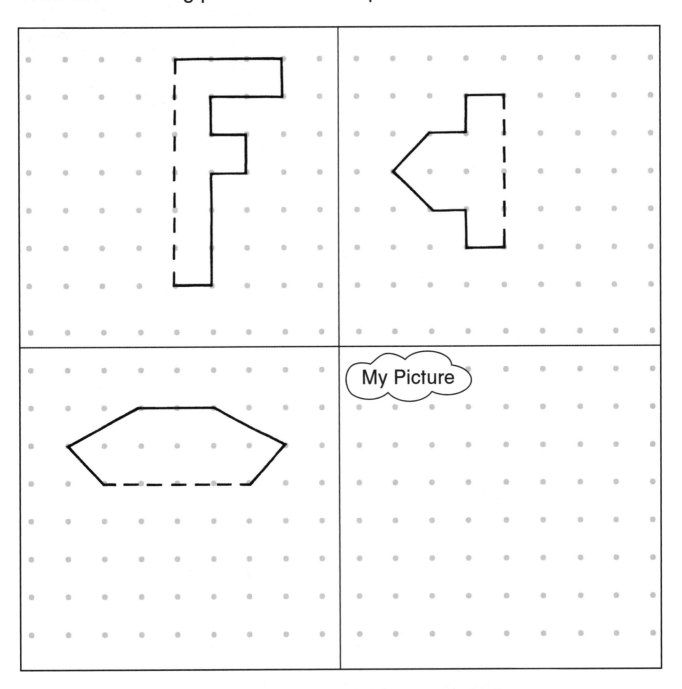

My Picture

Focus | Children draw a "mirror image" to complete each symmetrical figure.

We Found Symmetry!

Put a Mira on the square.
Look for matching parts.

Use  red .

Draw a line where you put the Mira.

Look for another way.

Use green .

Draw a line where you put the Mira.

How many ways can you find? _____

What did you find out?

Symmetry

Circle the pictures that show matching parts.
Use a Mira to help you.

Focus | Children use a Mira to determine which pictures show symmetry.

226 Unit 9, Lesson 5: Symmetry

Name: _____ Date: _____

Name: _____ Date: _____

Symmetry in Letters

Circle the letters that show symmetry.
Check using a Mira.

O G V C M

P A J B R X

Where would you place a Mira on MOM to show symmetry?
Use red .

Draw a line where you would put the Mira.

MOM

Think of another word that shows symmetry. _____
How can you tell it shows symmetry?

_____

HOME CONNECTION
Print your child's name in block letters. With your child, check if the name has symmetry. Look for symmetry in names of other family members.

Unit 9, Lesson 5: Symmetry **227**

Name: ______________________ Date: ______________________

Get Moving

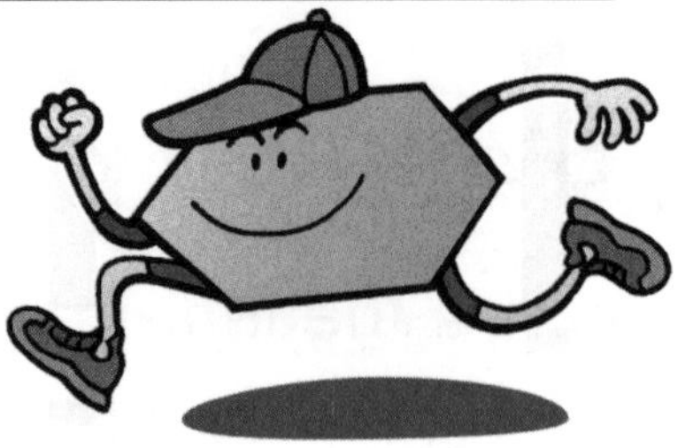

Place a cutout on .

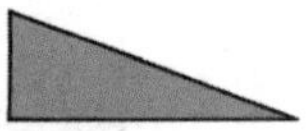

Move it to cover .

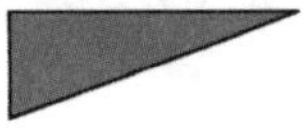

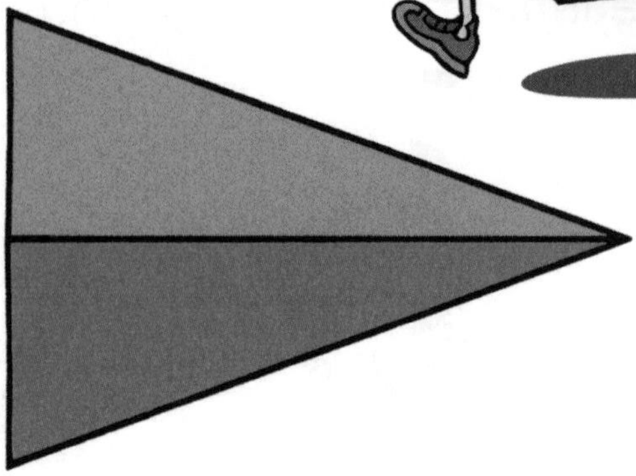

How did the triangle move?

____

Place a cutout on .

Move it to cover .

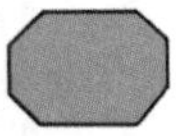

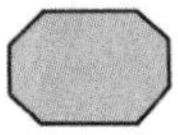

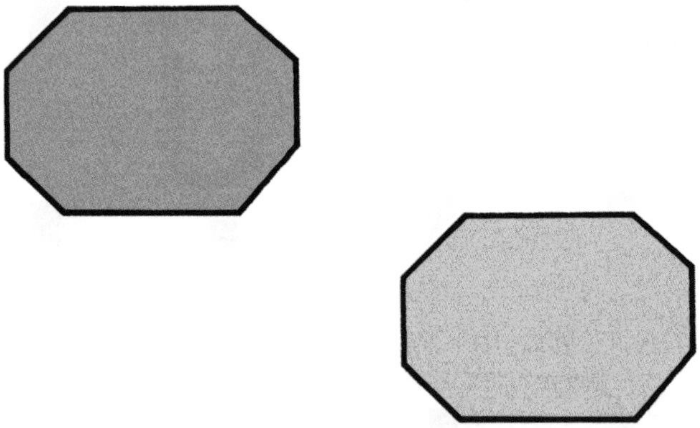

How did the octagon move?

____

Place a cutout on .

Move it to cover .

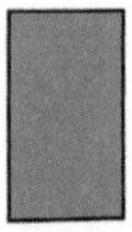

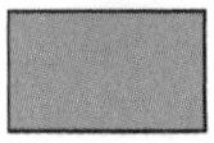

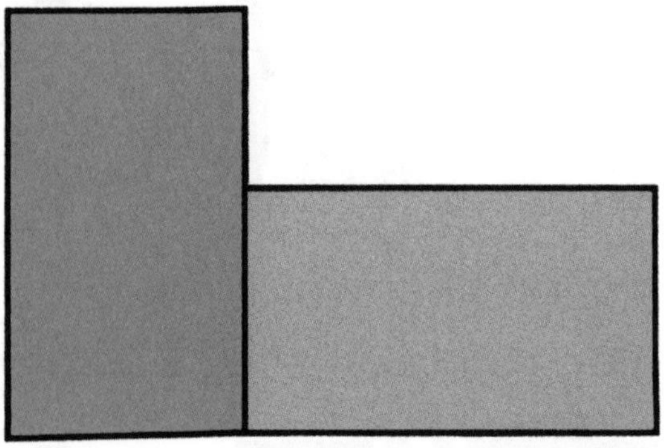

How did the rectangle move?

____

Focus | Children use cutouts from *LM 3: Geometric Figures* to move figures to new positions and describe how they did it.

Move on a 100-Chart

Start at 27.

Show one way to reach 54.

Use 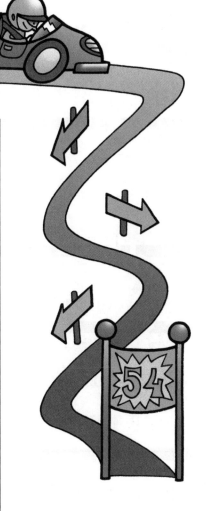 red .

Tell your partner how you did it.

1	2	3	4	5	6	7	8	9	10
11	12	13	14	15	16	17	18	19	20
21	22	23	24	25	26	27	28	29	30
31	32	33	34	35	36	37	38	39	40
41	42	43	44	45	46	47	48	49	50
51	52	53	54	55	56	57	58	59	60
61	62	63	64	65	66	67	68	69	70
71	72	73	74	75	76	77	78	79	80
81	82	83	84	85	86	87	88	89	90
91	92	93	94	95	96	97	98	99	100

Show another way to reach 54.

Use blue .

Tell your partner how you did it.

Focus | Children practise moving from one number to another on a 100-chart. They explain how they moved.

Find My Number

Choose a number on the 100-chart.
Don't tell!
Print directions to reach your number.

Start at _____.

1. _____

2. _____

3. _____

Ask your partner to find the number.

My partner thinks the number is _____.

Is your partner right? Circle Yes or No.

Try it again with a different number.

Start at _____.

1. _____

2. _____

3. _____

My partner thinks the number is _____.

Is your partner right? Circle Yes or No.

Focus | Children choose numbers on a 100-chart and write directions for others to find them.

Name: _____ Date: _____

Show the Way Home

Help the prairie dog find its way home.
Use  to show the way.

Print your directions for the prairie dog.

_____

_____

_____

Focus | Children draw a path and describe a way for the prairie dog to get home through a maze of tunnels.

HOME CONNECTION
Make a map of your home. Hide an object and then show where it is on the map. Ask your child to find it. Take turns hiding and following directions to find objects in your home.

A Quilt

Follow the directions on page 233 to complete the quilt.

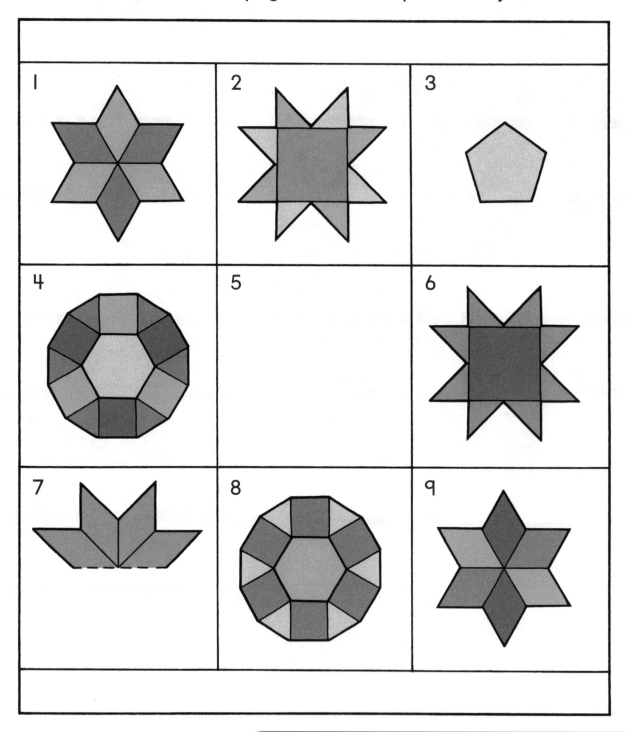

Focus | Children follow directions to complete a quilt.

HOME CONNECTION
With your child, take turns describing each of the quilt blocks on page 232. Ask your child to describe geometric designs in your home.

Finish the Quilt!

Part I

Draw the matching part to finish quilt block 7.

Find the quilt block that has only a pentagon.
Draw a triangle on each edge of the pentagon.

What does the design look like? _____

How many vertices are there in this block? _____

Use figures to make your own design in the centre quilt block.

Tell what you used. _____

_____

_____

Part 2

Draw a repeating pattern on the border at the top and bottom
of the quilt. Use just one figure. You can make it slide, flip, or turn.

Name two quilt blocks that have the same design.

_____ and _____

Start in Block I. Go straight down two blocks.
Go one block to the right. Go up one block. Go one block to the left.

What is the number of the block you are in? _____
Name the figures you can find in that block.

_____

Focus | Children follow directions to complete a quilt and answer related questions.

Unit 9, Lesson 8: Show What You Know **233**

Name: _____   Date: _____

My Journal

Tell what you learned about figures.
Use pictures, numbers, or words to show your thinking.

Focus | Children reflect on and record what they learned about 2-D geometric figures.

UNIT 10

Multiplication, Division, and Fractions

Focus | Children use skip counting and other methods to count items in the picture.

Name: _____ Date: _____

Dear Family,

In this unit, your child will build on number patterns to develop concepts of multiplication, division, and fractions, working with concrete materials.

The Learning Goals for this unit are to

- Learn about multiplication as counting groups of objects.
- Understand that repeated addition, skip counting, and multiplication are the same.
- Learn about division through grouping and sharing.
- Understand the meaning of halves, thirds, and fourths.
- Connect multiplying, dividing, and fractions to daily experiences with making equal groups and sharing.

You can help your child achieve these goals by doing the Home Connection activities suggested at the bottom of selected pages.

100-Chart Patterns

Shade every second number .
What is the skip counting pattern?

Count by _____.

Shade every fifth number 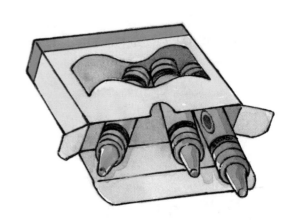.
What is the skip counting pattern?

Count by _____.

1	2	3	4	5	6	7	8	9	10
11	12	13	14	15	16	17	18	19	20
21	22	23	24	25	26	27	28	29	30
31	32	33	34	35	36	37	38	39	40
41	42	43	44	45	46	47	48	49	50
51	52	53	54	55	56	57	58	59	60
61	62	63	64	65	66	67	68	69	70
71	72	73	74	75	76	77	78	79	80
81	82	83	84	85	86	87	88	89	90
91	92	93	94	95	96	97	98	99	100

Show another skip counting pattern. What is your pattern?

_____

Focus | Children use colour to record skip counting patterns on a 100-chart.

Unit 10, Launch: Multiplication, Division, and Fractions

Name: _____ Date: _____

Button Up

Draw 4 groups of 2 buttons.

Write the addition sentence.

_____

4 groups of 2 buttons are

_____ buttons altogether.

Draw 3 groups of 4 buttons.

Write the addition sentence.

_____

3 groups of 4 buttons are

_____ buttons altogether.

Draw 5 groups of 3 buttons.

Write the addition sentence.

_____

5 groups of 3 buttons are

_____ buttons altogether.

Draw 3 groups of 6 buttons.

Write the addition sentence.

_____

3 groups of 6 buttons are

_____ buttons altogether.

HOME CONNECTION

Give your child objects in groups, such as 6 groups of
5 beans. Have your child show you how to find out how
many beans in all and to write a repeated addition sentence.

Focus | Children use repeated addition to count
groups of buttons.

Name: ______________________ Date: ______________________

Make a Problem

How many groups? _____

How many in each group? _____

Think of your problem. Use counters to solve it.

Draw to show how you solved it.

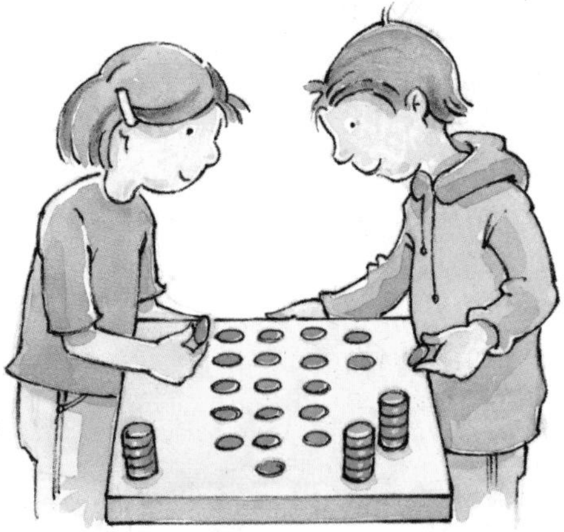

What is the addition sentence for your problem?

____

What is the multiplication sentence for your problem?

____

Focus | Children write and solve a problem using repeated addition and multiplication.

Multiply It!

How many items are in each box?

Write the addition sentence.

_____

Write a multiplication sentence.

_____

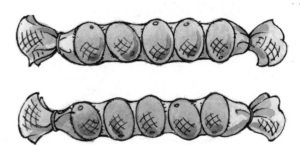

Write the addition sentence.

_____

Write a multiplication sentence.

_____

Write the addition sentence.

_____

Write a multiplication sentence.

_____

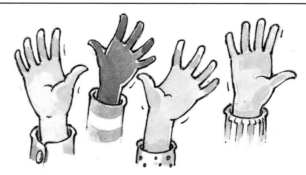

Write the addition sentence.

_____

Write a multiplication sentence.

_____

Focus | Children write about multiplication problems in two ways.

HOME CONNECTION

Find examples of items that come packaged in groups, such as stamps. Ask your child to write a repeated addition sentence and a multiplication sentence to find how many items there are altogether.

Share the Toys

How many toys are there altogether? _____

How many do you think each person will get? _____

Show how you divided the toys.

How many did each person get? _____

Write a sentence to show how the toys are shared.

_____

Share and Share Alike

Four children each want an equal share of apples to take home.
How many apples should each child get?

Write a sentence about your answer. _____

_____

Six children want to share these crayons fairly.
How many crayons should each child get?

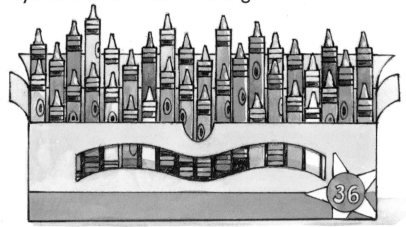

Write a sentence about your answer. _____

_____

HOME CONNECTION
Have your child find out how many pieces of fruit, such as
grapes, can be fairly divided among a group of friends or family.

Focus | Children practise dividing items into equal
groups (sharing) and recording the result.

How Many Bunches?

You are making bunches of flowers.
You have 36 flowers.

You put 6 flowers in every bunch.
How many bunches can you make?
Write a sentence about your answer.

_____

Suppose you put 4 flowers in every bunch.
How many bunches can you make?
Write a sentence about your answer.

_____

HOME CONNECTION
Have your child sort a number of objects (divisible by 4) into groups of 4. Ask your child how many groups he or she made.

FOCUS | Children divide 36 flowers into groups of 6 and of 4.

Name: _____   Date: _____

Parts of a Whole

Write if each figure is in halves, thirds, or fourths.

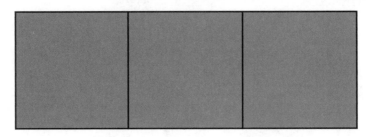

This is in _____. There are _____ equal parts.

_____

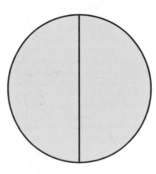

This is in _____. There are _____ equal parts.

_____

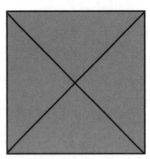

This is in _____. There are _____ equal parts.

Focus | Children describe fractional parts.

HOME CONNECTION
Let your child work with fractions during meals. Show your child a sandwich in halves, a pear cut into fourths, and so on. Have him or her name the fraction.

Colouring Fractions

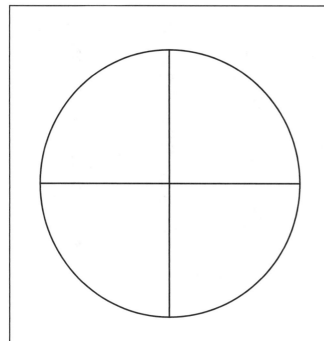

Colour one-fourth.

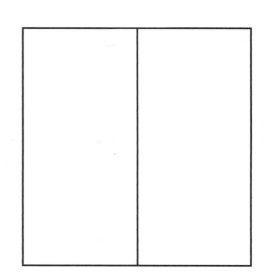

Colour one-half.

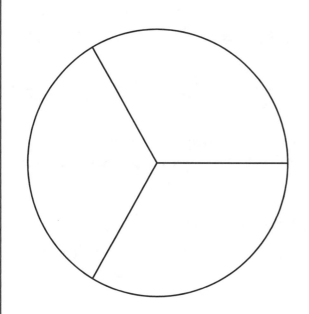

Colour two-thirds.

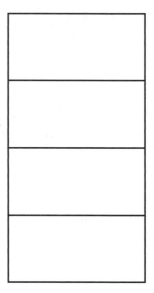

Colour three-fourths.

Tiny Toys

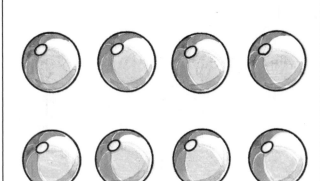

Show 2 equal groups.
How many balls in each half?

_____

Show 3 equal groups.
How many dolls in each third?

_____

Show 2 equal groups.
How many ducks in each half?

_____

Show 4 equal groups.
How many cars in each fourth?

_____

HOME CONNECTION

Find examples of items that come packaged in groups, such as rolls.
Ask your child to tell you how many are in one-half of the package.

Focus | Children determine fractions of a set.

Bumper Cars

There are 13 children lined up for the bumper cars.

There are 4 bumper cars.

Each bumper car can hold 3 children.

Can everyone ride bumper cars at the same time?

Show how you solved the problem.

Use pictures, numbers, or words.

Focus | Children use the strategy "act it out" to solve a problem, and record their work.

Getting on a Plane

Altogether, 22 people need to fly on this plane.

There are 6 rows of seats. There are 4 seats in each row.

Can everyone get on the plane?

Show your thinking in pictures, numbers, or words.

HOME CONNECTION

Invite your child to tell you how he or she solved the problem on this page and to make up a similar problem for you to solve. Use pennies to stand for the people who want to get on the plane.

Focus | Children solve a problem involving division concepts.

Musical Ride Problems

Look at the picture.

Tell a number story
about equal groups.
Show your story
and how you solved it.

```


```

How many groups? _____ How many in each group? _____

How many altogether? _____

Write a number sentence about your story. _____

Focus | Children create and solve a
story about equal groups.

Unit 10, Lesson 8: Show What You Know **249**

Name: ______________________ Date: ______________________

Equal Groups of Horses

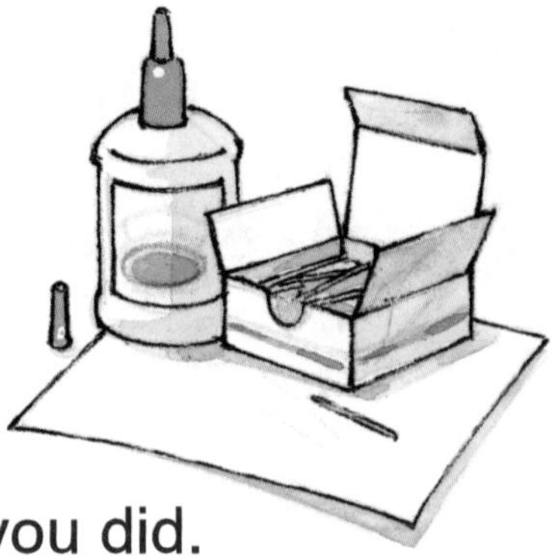

You need to put 24 horses in equal groups.
Use toothpicks to show equal groups.
Glue the toothpicks on a piece of paper.

Use pictures, numbers, or words to show what you did.

Focus | Children divide 24 into equal groups.

After the Ride

There are 24 horses.
One-third of the horses get new saddles.
Use counters to find the answer.

How many horses get new saddles?

There are 24 horses.
Four horses go in each trailer to travel home from the show.
Use counters to find the answer.

How many trailers do the horses need?

Four horses will eat equal shares of hay.
Fold a piece of paper to show how to divide the hay.

What fraction of the hay will each horse get? _____

Name: _____   Date: _____

My Journal

Tell what you learned about multiplying and dividing.
Use pictures, numbers, or words.

Tell what you learned about fractions.
Use pictures, numbers, or words.

Focus | Children reflect on and record what they learned about multiplication, division, and fractions.

Mass and Capacity

Focus | Children look for ways people are measuring and comparing items in the store, and identify measurable attributes.

253

Dear Family,

In this unit, your child will be learning more about measurement. Your child will develop an understanding of *capacity*, how much a container can hold, and *mass*, which relates to the heaviness of an object.

The Learning Goals for this unit are to

- Estimate, compare, measure, and order the capacities of containers by filling them with materials to see how much they hold.
- Estimate, compare, measure, and order the masses of objects using simple scales with non-standard units.
- Solve everyday problems about mass and capacity.

You can help your child achieve these goals by doing the Home Connection activities suggested at the bottom of selected pages.

What You Measure

Show things you measure.
Use pictures, numbers, or words.

Capacity

Home	School

Mass

Home	School

Name: _____   Date: _____

Comparing Capacity

Predict which container will hold the most.

_____

Predict which container will hold the least.

_____

Draw or write the name of each container.
Measure and record how many scoops fill each one.

Container A	Container B	Container C	Container D
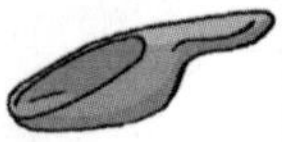 _____	 _____	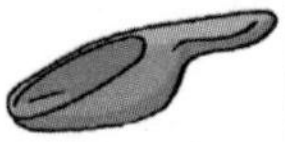 _____	 _____

Which container holds the most?

_____

Which container holds the least?

_____

List the containers in order from the greatest capacity to the
least capacity.

_____ _____ _____ _____

Focus | Children measure and compare the capacities of four containers.

Measure and Graph

In the chart, draw or write the name of each container.
Measure how many scoops fill each one.
Tally and record the number of scoops.

Container	Tally	Number of

Graph how many scoops fill each container.

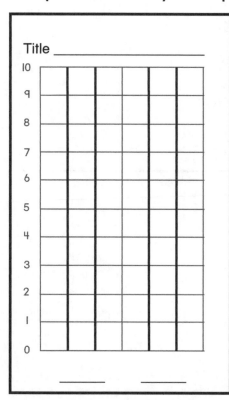

Title _____

10
9
8
7
6
5
4
3
2
1
0

_____ _____

What does the graph tell you?

_____

HOME CONNECTION

Gather some kitchen containers. Have your child compare two or more at a time, working over the sink, to see which holds more by pouring water from one into another.

Name: _____   Date: _____

About How Many Scoops?

My scoop looks like _____.

I estimate my container will hold about _____ scoops.

Fill your container half full. How many scoops do you need? _____

Change your estimate if you want. _____

Fill your container full. How many scoops does it hold? _____

Check with another group.

Their scoop looks like _____.

They needed _____ scoops.

What did you find out about estimating capacity?
Use pictures, numbers, or words.

FOCUS | Children estimate and check the number of scoops of material needed to fill a container. They compare their results with those of another group.

Ordering Capacity

In the chart, draw or write the name of each container.
Predict which one will hold the least.

_____.

Predict which one will hold the most.

_____.

Measure the capacity of each container.
Tally and record the number of scoops.

Container	Tally	Number of

List the containers in order from least capacity to greatest capacity.

_____   _____   _____

HOME CONNECTION

Enlist your child's help in everyday routines, such as making juice. Ask: "How many containers of water do we add? Is our pitcher big enough? Will it be full to the top when we are done?"

Focus | Children estimate the capacities of containers and measure the capacities using a scoop. Then they order the containers from least to greatest capacity.

Unit II, Lesson 2: Estimating Capacity **259**

Name: ______________________ Date: ______________________

Fair Shares to Drink

Estimate how many units of water will go in each glass. __________

Draw the glasses you are using to act out the problem.

____ units	____ units	____ units	____ units	____ units

Share the water among all 5 glasses.
Record the level of water in each glass.
Record the number of units in each glass.

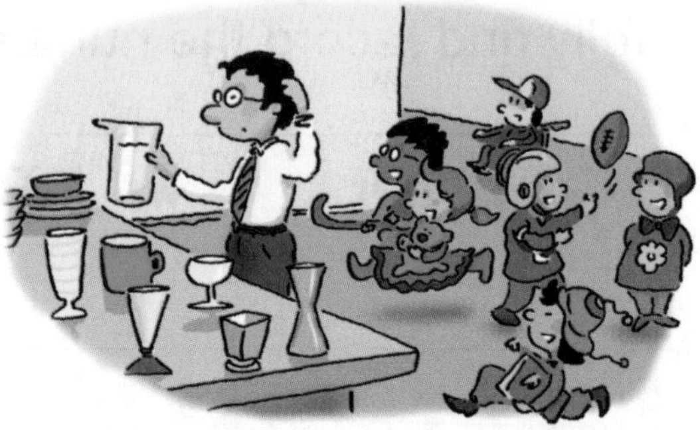

Explain how you solved the problem.
Use pictures, numbers, or words.

Focus | Children work with materials to share a given amount of water equally among 5 glasses and show the level for each glass when filled by an equal measure. Then they explain how they solved the problem.

Fair Shares to Snack

How many units do you think each bowl will get? _____

Draw the bowls you are using to act out the problem.

____ units	____ units	____ units	____ units	____ units

Share the popcorn among all 5 bowls.
Record the level in each bowl.
Record the number of units in each bowl.

Explain how you solved the problem.
Use pictures, numbers, or words.

HOME CONNECTION
With your child, divide treats equally among several containers. Have your child show the level for each container when filled by an equal measure.

Comparing Mass

Predict which object will be the lightest.
Predict which object will be the heaviest.

Use your predictions to order the objects from lightest to heaviest.
Draw or write the name of each object in order.

the lightest object			the heaviest object

Use a balance scale to compare the objects.
Line up the objects on your desk from lightest to heaviest.
List the objects in order from lightest to heaviest.

_____ _____ _____ _____

How did the balance scale help you? Use pictures, numbers, or words.

Focus | Children order the masses of items using their hands. They repeat the activity using a balance scale.

Name: _____   Date: _____

Classroom Objects

Choose 4 objects from around you.
Predict which object will be the lightest and which will be the heaviest.

Draw or write the name of each object.

the lightest object			the heaviest object

Use a balance scale to compare the objects.
List the objects in order from lightest to heaviest.

_____ _____ _____ _____

Now find another object to compare.
Do you think it is the heaviest, the lightest, or near the middle?

_____

Check with a balance scale.
List the 5 objects in order from greatest to least mass.

_____ _____ _____ _____ _____

Focus | Children order the masses of classroom items using a balance scale.

HOME CONNECTION
Have your child find objects lighter than, the same as, or heavier than an object. Check with a simple hanger scale. (Tie a plastic bag to each end of a hanger and balance the hanger on your finger.)

Balancing Act

Draw or write the name of each object
you are measuring.
Estimate the mass of each object.
Write each estimate in the chart.

Object	Estimated Mass	Measure
the lightest object		
the heaviest object		

Measure the mass of each object. Write each measure in the chart.

About How Many?

Estimate the masses of 2 objects.
Use 3 different units.

	Object 1	Object 2
Unit 1 estimate		
Unit 2 estimate		
Unit 3 estimate		

Use a balance scale.

Find the masses of the 2 objects.
Use the 3 different units.

	Object 1	Object 2
Unit 1 measure		
Unit 2 measure		
Unit 3 measure		

Compare with a friend.

Were your answers different?
Why do you think that happened?

_____

HOME CONNECTION

Have your child estimate masses at home. For example,
ask: "About how many lemons are as heavy as a grapefruit?
About how many grapes are as heavy as an orange?"

Focus | Children estimate and measure masses
using 3 different non-standard units.

Plant-Pot Parade

Estimate to order the pots from least
to greatest capacity.

_____ _____ _____ _____ _____

Draw or write the name of each plant pot in order.

Measure the capacity of each container.
Record the capacity beside its name or picture.

Which container holds the most?
How do you know?

_____

_____

Focus | Children estimate the capacities of containers and explain how to solve a capacity problem.

Name: ____________________ Date: ____________________

Which Is the Heaviest?

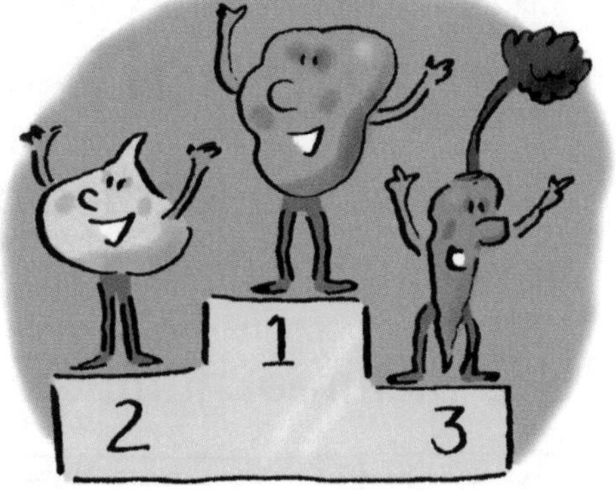

Estimate to order the objects from lightest to heaviest.

______ ______ ______ ______ ______

Draw or write the name of each object in order.

Measure the mass of each object.
Record the mass beside its name or picture.

Which object is the heaviest?
How do you know?

____

____

Focus | Children estimate and measure the masses of 5 objects to find the heaviest.

My Journal

Tell what you learned about comparing
how much different containers can hold.

Use pictures, numbers, or words to show your thinking.

Tell what you learned about comparing how heavy objects are.

Use pictures, numbers, or words to show your thinking.

Focus | Children reflect on and record what they learned about capacity and mass.

HOME CONNECTION
Invite your child to practise measuring capacity and mass while helping with grocery shopping or preparing simple recipes.

The Field Trip

Aquarium Place

The Grade 2 class was so excited.
Even Grandma was invited.
Field Trip Day was finally here.
Everyone would visit Aquarium Place this year.

Inside they hardly made a sound.
There were glass and water tanks all around.
And so many different fish were there.
All colours and sizes—they were everywhere!

A whale swam by and looked their way.
"Welcome," its expression seemed to say.
Over in the corner they saw a crowd.
"A shark!" they suddenly gasped out loud.

Then some scuba divers came
and swam with fish that seemed quite tame.
"That looks like Grandma!" said Cam's friend Ben.
Cam blinked his eyes and looked again.

"I wonder," said Cam as he stopped to stare.
"Just how many fish might fit in there?
How many tubs of water, too?"
"I'll bet there are more than one thousand!" cried Lu.

Soon it was time for the dolphins to eat.
Buckets of fish were the lunchtime treat.
Some helpers held fish out over the tanks,
and dolphins jumped up and chirped out, "Thanks!"

What an amazing day it had been.
There were all kinds of undersea life they had seen.
They lined up again to get back on the bus, as they
wondered, "What did the fish think when they looked at *us*!"

About the Story

The story was read in class to prepare for a Mathematics Investigation activity. Children used positional language, solved number and measurement problems, and made a model using 3-D solids.

Talk about It Together

- What is your favourite part of the story?
- What is Grandma doing while the children look at all the fish in the tanks?
- Do you think working at Aquarium Place would be interesting? Why? Why not?
- What kind of jobs would the aquarium workers do?

At the Library

Ask your local librarian about other good books to share about measuring objects, mass and capacity, geometric solids, and numbers.

Where Will You Be?

Put X's on the map on 3 things you want to see.
Draw a line showing how to get from one to another.
Write a note telling where you will be.

Dear Teacher,
First, I will visit _____
Follow these directions to find me.

_____

_____

Second, I will visit _____.
Follow these directions to find me.

_____

_____

Third, I will visit _____.
Follow these directions to find me.

_____

_____

How Big Is a Whale?

Fun Facts about Orca Whales

An orca whale can be 7 m long.

An orca whale can be 2 m wide.

An orca whale can eat 18 buckets of fish every day.

An orca whale can have the same mass as 150 Grade 2 children!

Look at the fun facts about orca whales.
About how many children could lie in a row beside one whale? Show your thinking.

Estimate how many whales could fit in your classroom.
Use pictures, numbers, or words to show how you could check your estimate.

Create a whale measurement problem of your own and solve it.

Moving to a New Home

There are 48 fish in the tank.

Place the fish equally into 3 new tanks.

How many will be in each tank? _____
Use counters to help you solve the problem.
Show your thinking in pictures, numbers, or words.

Place the fish equally into 8 new tanks.

How many will be in each tank? _____
Use counters to help you solve the problem.
Show your thinking in pictures, numbers, or words.

Building the Aquarium

Here is the outside of an aquarium.

Use 3-D solids to make a model of an aquarium building.
What solids did you use?

_____

_____

Use pictures, numbers, or words to tell how you made
your model.

Symmetrical Art

Fold a piece of paper in half.
Then open it and put a few
dabs of paint on one half.
Fold the paper again and
press the halves together.

Open it and you will see a
symmetrical masterpiece!

Want to make it more interesting?
Cut out some figures and glue them on your artwork.
Remember, whenever you put something on one side,
it must have a symmetrical match on the other side!

Eggs for All

Take an empty egg carton
and put a counter
in each space.
Let's imagine brunch
is at your place.
Each guest eats the same
number of eggs, and all
the eggs are used.

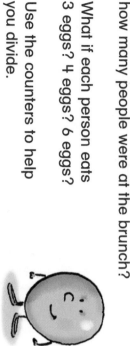

If each person eats 2 eggs,
how many people were at the brunch?

What if each person eats
3 eggs? 4 eggs? 6 eggs?

Use the counters to help
you divide.

The next 4 pages fold in half to make an 8-page booklet.

Fold

Math at Home

I love math
and math loves me!
It's not too hard to see.
For everywhere I seem to go,
math always follows me!

My cookie has 12 chocolate chips.
Its shape is nice and round.
I split it fairly into halves.
See? Math *is* all around!

Math at Home 3

Tell Me About This Figure
Game Board

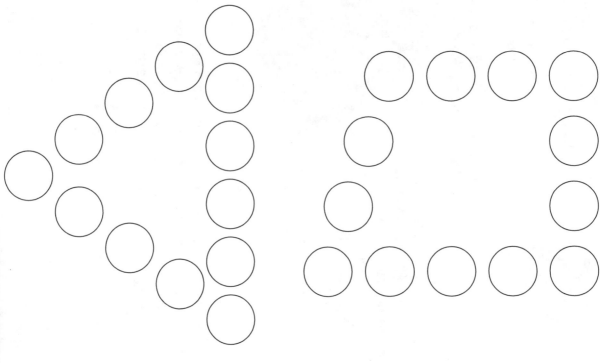

No Fair

Wendy's little brother
thought she had more
juice than he did.
She tried to tell him
that she really had less.
What do you think
she said to him?

Sandwich Fractions

Get creative with your lunch.
Cut your sandwich a different way each day.
You might try

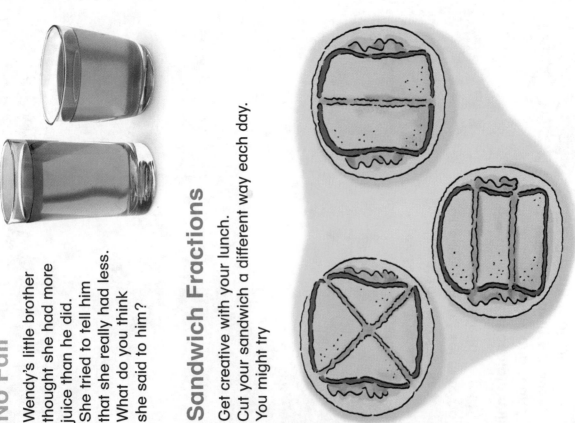

The sky's the limit!

Tell Me About This Figure

You'll need:

- 2-D figures in a bag you cannot see through

- small counters
- game board (page 7)

Before the game begins, each player chooses an outlined figure on the game board (page 7).

On your turn:

- Pull a figure out from the bag and lay it between you and your partner.
- Both of you secretly write one thing that you notice about the figure. (Think about number of sides, lengths of the sides, number of corners.)
- Read your descriptions to each other.
  If they are **different**, you place two counters on your outlined figure.
  If they are the **same**, the other player places one counter on his or her outlined figure.

Take turns until someone's outlined figure is full.

Moving Day

Ryan could move this box:

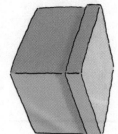

But he could not move this one:

What do you think is in each box?

Stick Figures

Suppose it takes 3 craft sticks to make one side of a square.

How many craft sticks will it take to make the whole square?

Suppose it takes 3 craft sticks to make one side of a hexagon.
How many craft sticks might it take to make the whole hexagon?

What other ways are there to make the whole hexagon?

Amazing Area

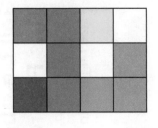

This painting
is made completely
of squares.

It has an area of _____ squares.

Imagine the artist wants a frame.
The distance all the way around is _____ units.
(Count each side of a square as one unit.)

Let's imagine the artist makes another painting using
the same squares.

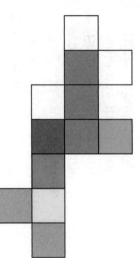

What happens
to the area
of the painting?

What about the
distance around?

How Much Is 100 Drops?

Imagine putting 100 drops
of water in a drinking glass.
How full will it be?

Take a guess, then try it!
Were you surprised?

Transformation Moves

Challenge a friend to follow your directions.

You might say: Turn to the left.
                Slide to the right.
                Turn to the right.
                Slide forward.

Get creative! Switch roles and see
how much fun directions can be.

How Long Is It?

Find something in your house
that is about

- 3 footprints long
- 4 fingers wide
- 2 arms long
- 5 hands wide

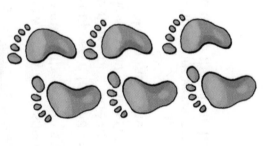

Guess first, then measure
the object.

What do you think would
happen if a grown-up looked for
something 4 footprints long?